—

Invisible

Invisible

The Dangerous Allure of the Unseen

Philip Ball

The University of Chicago Press

CHICAGO

PHILIP BALL is a freelance writer who lives in London. His many books include *Curiosity: How Science Became Interested in Everything* and *Serving the Reich: The Struggle for the Soul of Physics under Hitler*, both also published by the University of Chicago Press.

The University of Chicago Press, Chicago 60637
© 2015 by Philip Ball
All rights reserved. Published 2015.
Printed in the United States of America

24 23 22 21 20 19 18 17 16 15   1 2 3 4 5

ISBN-13: 978-0-226-23889-0 (cloth)
ISBN-13: 978-0-226-23892-0 (e-book)
DOI: 10.7208/chicago/9780226238920.001.0001

Originally published by Bodley Head, 2013.

LIBRARY OF CONGRESS CATALOGING-IN-PUBLICATION DATA
Ball, Philip, 1962– author.
  Invisible : the dangerous allure of the unseen / Philip Ball.
    pages : illustrations ; cm
  Includes bibliographical references and index.
  ISBN 978-0-226-23889-0 (cloth : alk. paper) —
  ISBN 978-0-226-23892-0 (e-book) 1. Invisibility. I. Title.
  QC406.B35 2015
  535'.1—dc23

                                              2014035709

Contents

Illustrations

Images: (left) Hans-W. Ackermann, Department of Microbiology, Medical School, Laval University, Quebec, Canada; (centre) Centers for Disease Control and Prevention/Dr Fred Murphy and Sylvia Whitfield; (right) Reo Kometani and Shinji Matsui, University of Hyogo.

p.221   *a*: The microscopic 'pond life' of London, from *Punch*, 11 May 1850. *b*: The cholera microbe in the Parisian *Le Grelot*, 1884. *c*: Poster from a Parisian theatre in 1883. *d*: 'Germs' depicted in current teaching resources by the Government of Western Australia Departments of Health and Education.

p.224   An artist's impression of a nanobot removing plaque from the walls of a human blood vessel. Image: Hybrid Medical Animation/Science Photo Library.

p.225   *Fantastic Voyage* (1966).

p.226   A bug on microscopic gears of a microelectromechanical device carved into a silicon chip. Image: courtesy of Sandia National Laboratories, SUMMiT™ Technologies, www.mems.sandia. gov.

p.227   The Michigan Micro Mote.

p.230   The 'invisibility cloaks' of Susumu Tachi. Images: courtesy of Susumu Tachi, University of Tokyo.

p.230   Liu Bolin painted to vanish into the background. Images: courtesy of Eli Klein Fine Art © Liu Bolin.

p.231   An artist's impression of Tower Infinity in South Korea. Images: GDS Architects.

p.233   The chameleon. Image: Ales Kocourek.

p.233   A flatfish, almost invisible against the sandy sea bed. Image: Moondigger.

p.235   Stacked plates of the reflectin protein in iridophore cells create tunable reflective colours in squid. Image: courtesy of Rajesh Naik, Wright-Patterson Air Force Base.

p.235   A moth camouflaged against tree bark. Image: © Marc Parsons/ Dreamstime.com.

p.237   The zebra may be hidden among light-coloured vegetation, but in grassland the stripes have no camouflaging effect. Images: (left) Rei; (right) Gusjer.

p.238   'Differential blending' of a patchy or mottled animal, from Hugh Cott's *Adaptive Coloration in Animals*.

Preface

I had hoped to find occasion somewhere in this book to mention the delightful little book by Yoshi Oida called *The Invisible Actor*. I never did. That is because, I now realise, it belongs here in the preliminaries, where the author stands on the stage in person and prepares to disappear. Like the actor, he or she must subsequently become invisible, and what remains is a performance.

Oida has many thoughtful things to say about performance. There are lessons in his book for the writer but I don't think I have yet fully understood them, much less mastered them. In the Kabuki theatre, he says,

> there is a gesture which indicates 'looking at the moon', where the actor points into the sky with his index finger. One actor, who was very talented, performed this gesture with grace and elegance. The audience thought: 'Oh, his movement is so beautiful!' They enjoyed the beauty of his performance, and the technical mastery he displayed.
>
> Another actor made the same gesture, pointing at the moon. The audience didn't notice whether or not he moved elegantly; they simply saw the moon. I prefer this kind of actor: the one who shows the moon to the audience. The actor who can become invisible.

This applies to fiction writers: there are those we admire for their style, and there are those we admire because they vanish so that all we see is the story. The former may be consummate performers, but the latter are magicians. And we might wonder, as many novelists wonder of, say, Penelope Fitzgerald, 'how did they *do* that?'

How does this notion translate to non-fiction? All I can say for sure is that the performative aspect remains, and that what Oida says of performance remains true also: 'Your job as an actor is not to display how well you perform, but rather through your performance to enable the stage to come alive.'

What is also true, however, is that the performer doesn't achieve any of this unaided. There are directors, technicians and engineers, back-of-house staff. There's no show without them. I am fortunate indeed to have such a dependable and experienced crew: my agent Clare Alexander and my editors and copy-editor at Bodley Head, Katherine Ailes, Jörg Hensgen and Kay Peddle. I have also benefitted from a generous and deeply knowledgeable team of advisers: for their thoughtful comments, corrections and general conversation I am grateful to Ruth Bottigheimer, William Brock, Huanyang Chen, Owen Davies, Claire Hardaker, Ulf Leonhardt, Richard Noakes, John Pendry, Roberto Piazza, Christopher Priest, Hollis Robbins, James Russell, David Smith and Francisco Vaz da Silva. And for posing an innocent question that set the whole performance in train, I owe a great debt to Anais Tondeur. My family don't need to see the performance, because they have to live every day with the rehearsals. They are the reason I get through them.

Philip Ball
London, May 2014

1 Why We Disappear

It seemed that the ring he had was a magic ring: it made you invisible! He had heard of such things, of course, in old tales; but it was hard to believe that he really had found one, by accident.

J. R. R. Tolkien
The Hobbit (1937)

And perhaps in this is the whole difference; perhaps all the wisdom, and all truth, and all sincerity, are just compressed into that inappreciable moment of time in which we step over the threshold of the invisible.

Joseph Conrad
Heart of Darkness (1899)

In old tales – and usually in our new ones too – no one becomes invisible without a motive. It's the peculiarity of our times that we focus on the means and not the motive. Scientists and technologists today are slowly finding out how to build what they like to call invisibility cloaks and the world watches, for the most part entertained and amazed. But in the old stories, in myths and legends and fairy tales, invisibility was never so laboriously achieved, nor so compromised in the achievement. Making something invisible demanded special knowledge or special favours, but once that ability was secured, the magic simply happened. No one was particularly surprised or impressed by the feat itself; what mattered was not how but why you did it.

What is so easily forgotten when legend and fable are enlisted as a charming bit of scene-setting for announcements of technological advance is that these stories are not engineering challenges set by our ancestors. They might be filled with gods and devils, imps and giants, but they are really about our own world and the things we do to one another. It's in this sense that we have always possessed the secret of invisibility, and have always known where it might lead. We know what powers it conveys, and what dangers.

These are the subjects of my book, and this is why – more than for any banal chronological purposes – it must start at the beginning. For in the history of invisibility, the punchline comes at the outset: it is the earliest manifestations that tell us, in some respects, all we need to know about invisibility. The rest is 'just' the engineering. But it is the engineering – the 'how can we do this' – that discloses most eloquently the complications and repercussions that appear when myth collides with reality. In the gap between what we hoped for and what we got is a glimpse of who we are.

The magic ring

If you could be invisible, what would *you* do? The chances are that it will have something to do with power, wealth or sex. Perhaps all three, given the opportunity.

If that's so, there is no need to feel guilty. Or rather, it is doubtless good for the soul to experience a little contrition, but your response is not perverse or aberrant. We have it on Plato's authority that this is all perfectly normal. In the *Republic* he (or rather, his narrator Glaucon) explains that invisibility is not a technical problem but a moral one.

There are several accounts of how Gyges, the ancestor of King Croesus of Lydia, rose from humble origins to found the third dynasty of Lydian kings in the first millennium BC. All of them present him as a usurper and several say he was driven by lust both carnal and political. Gyges, it is generally agreed, stole from Candaules of Lydia both his throne and his wife. According to Herodotus, the old king brought it upon himself by ordering Gyges, who was his bodyguard at the time, to look secretly upon his queen so that he might be compelled to admit her outstanding beauty.* Gyges complied unwillingly,

* This kind of voyeurism whereby one puts one's partner on sexualized display is now called candaulism.

but the queen discovered him in his hiding place and, enraged by her husband's shameful behaviour, gave Gyges the option of killing the king or being put to death himself. He could hardly be blamed for the choice he made.

But Plato's account does not offer these extenuating circumstances. His Gyges begins as a shepherd in Candaules' service. While Gyges was tending his flock one day, an earthquake split the ground apart and he descended into the crevice. There he saw a horse made of bronze with doors in its side and, opening them, the dead body of a naked man lying inside, with a gold ring on his finger. Gyges took the ring and put it on.

After returning to the world above, Gyges met with his fellow shepherds, as was their custom, to prepare a monthly report on their flocks for the king. While sitting among his colleagues he happened to turn the ring's collet (the broad flange where a gem may be set) towards his palm, whereupon he vanished from the sight of the company. When he turned the collet outwards, he became visible once again.

That was all it took for Gyges to hatch a bold and mendacious scheme. He contrived to be made one of the messengers who delivered the report to the king, whereupon Plato's version lurches precipitously from bucolic fable to Sophoclean tragedy. As soon as he arrived, Plato writes, Gyges 'committed adultery with the king's wife, attacked the king with her help, killed him, and took over the kingdom'. These crimes, we are clearly meant to infer, were all done with the aid of Gyges' ring of invisibility.

The moral of the tale, says Glaucon, is that with such a magical charm, no one

> would be so incorruptible that he would stay on the path of justice or bring himself to keep away from other people's property and not touch it, when he could with impunity take whatever he wanted from the market, go into houses and have sexual relations with anyone he wanted, kill anyone, free all those he wished from prison, and do the other things which would make him like a god among men.

Don't imagine that Plato sees this as an unnatural or particularly reprehensible reaction. Glaucon admits that it would be naive to expect anything other than abuse of the privilege of invisibility:

The man who did not wish to do wrong with that opportunity, and did not touch other people's property, would be thought by those who knew it to be very foolish and miserable. They would praise him in public, thus deceiving one another, for fear of being wronged [themselves].

The problems that this poses for the rectitude of state authority – where 'those who practice justice do so against their will because they lack the power to do wrong' – occupy much of the remainder of the *Republic*.

For Plato, then, invisibility was not a wondrous power but a moral challenge – to which none of us is likely to prove equal. Invisibility corrupts; nothing good could come of it. In particular, invisibility will tempt us towards three things: power, sex and murder. This is the promise that has lured people to seek invisibility throughout time, whether by magical spells or esoteric arts or devices and garments that confer the ability to vanish.

The erotic unseen

Concealment was a useful attribute in the ancient world, where hazards could befall you anywhere. In early Christianity a magical power like invisibility was apt to be denounced as witchcraft – the occasional instances of magic in the Bible are depicted as deceitful trickery. But invisibility was sometimes permitted to saints, who, often originating through a pious retelling of local folklore, enjoyed a latitude not afforded to characters in the Scriptures. The *Dialogues* of Pope Gregory the Great in the sixth century are full of these dubious miracles: a monk, for example, becomes invisible when a group of Franks arrives to plunder his riches. Saint Patrick is said to have eluded the Druidic wizards of Ireland with an invisibility spell.

In mythic and traditional tales, invisibility is almost never a 'power of the body'. It is not that the unseen person knows how to turn him or herself invisible, but rather, this magical advantage is conferred by a talisman of some kind, an object that must be worn. This is more an act of concealment than of vanishing. Very often the talisman is a cap or a cloak – if indeed the two garments sometimes seem almost interchangeable, that is at least partly a linguistic quirk, because in tales of Germanic origin *Kap* (cape) could easily be conflated with *Kappe* (cap).

Athena gave Perseus a cap or helmet of invisibility that enabled him to escape from the Gorgons after he had slain their sister Medusa; the goddess wore it herself to fight Ares during the Trojan Wars.* The Tarnhelm of Richard Wagner's *Ring Cycle*, a magical helmet that enables one to change appearance as well as to become invisible, seems to be the composer's own invention, for there is no such object in the original legend of the Nibelungs. But, being made by the brother of the dwarf smith Alberich, it can claim some mythical precedent in the dwarfish *Huliðshjálmr* or 'concealing helmet' that features in some Norse tales.

A cloak of invisibility is used by an old soldier to follow the 'twelve dancing princesses' in the Grimms' fairy tale of that name, whereby he discovers the reason why their dancing shoes get worn out during the night: they are secretly going off to dance with twelve dashing princes. For solving this mystery, the soldier is granted the hand in marriage of the eldest daughter and so becomes heir to the throne. That invisibility is here the route to royal power echoes the Gyges story – but one can't fail to notice the recurring element of sexual voyeurism (we can hardly doubt that 'dancing' here is a euphemism), and the gift of invisibility is again freighted with erotic potential.

Invisibility in myth is often tied to sex and seduction. In the *Iliad*, Zeus cloaks Hera in a 'golden cloud' (unseen often blurs into unseeable) so that they might lie together on top of Mount Ida without the other gods spying on them. A magic ring enables Owain to seduce the Lady of the Fountain in the Welsh *Mabinogion*. The Egyptian sage Nectanebus used his powers of invisibility to deceive the Macedonian king Philip and his wife Olympias so that he might father Alexander the Great by the queen. According to folklore expert Francisco Vaz da Silva, 'cloaks and rings of invisibility are used mostly to enter an

---

* Homer calls it the Helmet of Hades and this same 'helmet of Pluto' later became a synonym for invisibility. It isn't clear whether this association with the god of the underworld comes from anything more than a linguistic elision: Greek *aidos* can mean invisible, although more commonly it connotes shame or modesty. But the attribution of invisibility as a power of the underworld chimes with the story of how Gyges found his ring and also with the general notion of being hidden from sight.

otherworldly realm where the protagonist will either seduce or deliver a princess, or take back his enchanted beloved.'

That's how the eponymous hero uses his invisibility cloak in the Italian fairy tale *Lionbruno*. The lad is betrothed to the fairy queen Madonna Aquilina after she rescues him from being pledged to the devil by his father's Faustian pact. But his subsequent misdemeanours infuriate Madonna Aquilina, and she banishes him from the fairy kingdom until he has 'worn out seven pairs of iron shoes'. Wandering as a despondent pilgrim, Lionbruno takes a cloak from a band of robbers encountered in a forest (that ubiquitous place of enchantment), tricking them into letting him try it on and so escaping from them. Concealed thus, he is borne by the Sirocco wind back to the fairy realm, where he climbs unseen through the queen's window and hides under her bed. After playing a jest on her by eating her supper (or in the more erotically explicit early versions, by kissing her) while still invisible, he reveals himself and they are reconciled. As one version has it, 'they threw their arms about each other with the truest love, and upon that bed they made their peace.'

Invisibility, then, provides access to liminal places tinged with desire, allure and possibility. Such allegorical content means that magical invisibility in fiction should never function simply as a convenient power that advances the narrative. It should not be bought cheaply, nor used idly. That is why the One Ring in *The Lord of the Rings* supplies a more satisfying, more mythically valid emblem than the cloaks of invisibility in the Harry Potter series. The latter, made from the hair of a creature from the Far East that can make itself invisible, are trinkets, a piece of incidental, even mundane magic. But magic must not be incidental or mundane, for it pulls on a subtle web of forces and must therefore have consequences.* Frodo Baggins' ring will, in the end, steal souls and reduce the bearer to a pitiful, malevolent wraith. This is what invisibility, when depicted in its truthful symbolic guises, does to us: it transforms us and pulls us into another realm. Even if that offers some immediate advantage, we had better

---

* I have a little more time for the Cloak of Invisibility that constitutes one of the Deathly Hallows, since this offers the more profound power of being able to hide from Death. It was in fact Death who awarded the cloak in the first place to one of three Peverell brothers who cheated him. But Harry Potter uses this garment in too many trivial ways for this potentially potent symbolic function to carry much weight.

not stay this way for too long. Invisibility is a state in which we mustn't linger or be trapped. The 'invisible child' in Tove Jansson's short story of that name in *Tales from Moominvalley* has faded into her unseen state through neglect and cruelty, and needs to be coaxed back to visibility with love, making this one of the few modern children's stories wise enough to avoid suggesting that this is a 'superpower' it would be fun to possess.

Invisible children

Both the antiquity of speculations about invisibility and its ubiquity as a trope of children's tales should come as no surprise, because a belief in one's ability to become 'unseen' seems to be an innate and normal part of the child's mental landscape. Invisible friends and pets give solace to most children at some point, and children up to about the age of four can disappear at will (so they insist) simply by closing or masking their eyes. As is so often the case with the ways of children, understanding this apparently puerile irrationality seems likely to cast some light on our own cognitive processes. Psychologist James Russell and his coworkers say that children undergo a developmental period 'in which they believe the self is something that must be mutually experienced for it to be perceived'. One might read this as a more general statement about social visibility and its absence.

The child's belief in her own invisibility with closed eyes turns out to be an epistemologically complex statement. The child does not exactly think that her body is hidden from view: whether or not she can be *seen* is a different matter from whether or not her body is visible. This subtle relationship between body and self becomes clear when Russell and his colleagues tested children aged between two-and-a-half and four by placing masks over the children's eyes and asking 'Can I see you?' In that situation the children would generally say 'no'. But if asked 'Can I see your head?', they would typically reply in the affirmative. They gave the same responses in relation to a third person whose eyes were masked:

'Can I see them?' – 'No'.
'Can I see their head?' – 'Yes'.

Further testing suggested that, for children, the act of seeing a person – which is to say, of *knowing about the person's presence* – depends on a mutuality of gaze: the child believes that only when an observer locks eyes with her can he register her actuality. To put it another way, for the *person* to be seen, it is not enough for the *body* to be visible: seeing is 'eyes meeting'. In this way, a person's visibility becomes, to a child, both a matter of choice and a situation that is socially defined: it requires the consent of both sets of eyes. One wonders what this says about the self-image of visibility of the child who assiduously avoids eye contact, as in some forms of autism.

This is a disconcerting, almost dizzying thought: we're left thinking not how a child can be so foolish as to imagine that they vanish by hiding their eyes, but rather, how extraordinary it is that the self is not located from birth in the physical body – that we have to learn to put it there. Even in maturity we do this only partially and conditionally: there is still a self that we don't wholly equate with the body. 'Do you like it?', I might ask, and you don't think for a moment that I am asking 'Does your body like it?' In this sense the self is always immaterial and unseen; but we learn to accept that it is shackled to visible flesh and blood.

Looked at this way, vanishing in fairy tales – whether to hide, to spy, or to wreak mischief – is not an extraordinary power at all, at least as understood by the youngest children. It is a power that we all have, but that we must relinquish along with infancy.

And we do. But the dream and the desire remain.

2 Occult Forces

Then charm me, that I
May be invisible, to do what I please
Unseen of any . . .

Christopher Marlowe
Dr Faustus

To render oneself invisible is a very easy matter, [but] it is not altogether permissible, because that by such a means we can annoy our neighbour in his (daily) life . . . and we can also work an infinitude of evils.

The Book of the Sacred Magic of Abramelin the Mage
Ed. S. Liddell Macgregor Mathers (1898)

Around 1680 the English writer John Aubrey recorded a spell of invisibility that seems plucked from a (particularly grim) fairy tale. On a Wednesday morning before sunrise, one must bury the severed head of a man who has committed suicide, along with seven black beans. Water the beans for seven days with good brandy, after which a spirit will appear to tend the beans and the buried head. The next day the beans will sprout, and you must persuade a small girl to pick and shell them. One of these beans, placed in the mouth, will make you invisible.

This was tried, Aubrey says, by two Jewish merchants in London, who couldn't acquire the head of a suicide victim and so used instead that of a poor cat killed ritualistically. They planted it with the beans in the garden of a gentleman named Wyld Clark, with his permission. Aubrey's deadpan relish at the bathetic outcome suggests he was sceptical all along – for he explains that Clark's rooster dug up the beans and ate them without consequence.

Despite the risk of such prosaic setbacks, the magical texts of the Middle Ages and the early Enlightenment exude confidence in their prescriptions, however bizarre they might be. Of course the magic will work, if you are bold enough to take the chance. This was not merely a sales pitch. The efficacy of magic was universally believed in those days. The common folk feared it and yearned for it, the clergy condemned it, and the intellectuals and philosophers, and a good many charlatans and tricksters, hinted that they knew how to do it.

It is among these fanciful recipes that the quest begins for the origins of invisibility as both a theoretical possibility and a practical technology in the real world. Making things invisible was a kind of magic – but what exactly did that mean?

Historians are confronted with the puzzle of why the tradition of magic lasted so long and laid roots so deep, when it is manifestly impotent. Some of that tenacity is understandable enough. The persistence of magical medicines, for example, isn't so much of a mystery given that in earlier ages there were no more effective alternatives and that medical cause and effect has always been difficult to establish – people do sometimes get better, and who is to say why? Alchemy, meanwhile, could be sustained by trickery, although that does not solely or even primarily account for its longevity as a practical art: alchemists made much else besides gold and even their gold-making recipes could sometimes change the appearance of metals in ways that might have suggested they were on the right track. As for astrology, it's persistence even today testifies in part to how readily it can be placed beyond the reach of any attempts at falsification.

But how do you fake invisibility? Either you can see something or someone, or you can't.

Well, one might think so. But that isn't the case at all. Magicians have always possessed the power of invisibility. What has changed is the story they tell about how it is done. What has changed far less, however, is our reasons for *wishing* it to be done and our willingness to believe that it can be. In this respect, invisibility supplies one of the most eloquent testimonies to our changing view of magic – not, as some rationalists might insist, a change from credulous acceptance to hard-headed dismissal, but something far more interesting.

How to be invisible

Let's begin with some recipes. Here is a small selection from what was doubtless once a much more diverse set of options, many of which are now lost. It should give you some intimation of what was required.

John Aubrey provides another prescription, somewhat tamer than the previous one and allegedly from a Rosicrucian source (we'll see why later):

> Take on Midsummer night, at xii [midnight], Astrologically, when all the Planets are above the earth, a Serpent, and kill him, and skinne him: and dry it in the shade, and bring it to a powder. Hold it in your hand and you will be invisible.

If it is black cats you want, look to the notorious *Grand Grimoire*. Like many magical books, this is a fabrication of the eighteenth century (or perhaps even later), validated by an ostentatious pseudo-history. The author is said to be one 'Alibeck the Egyptian', who allegedly wrote the following recipe in 1522:

> Take a black cat, and a new pot, a mirror, a lighter, coal and tinder. Gather water from a fountain at the strike of midnight. Then you light your fire, and put the cat in the pot. Hold the cover with your left hand without moving or looking behind you, no matter what noises you may hear. After having made it boil 24 hours, put the boiled cat on a new dish. Take the meat and throw it over your left shoulder, saying these words: "accipe quod tibi do, et nihil ampliùs." [Accept my offering, and don't delay.] Then put the bones one by one under the teeth on the left side, while looking at yourself in the mirror; and if they do not work, throw them away, repeating the same words each time until you find the right bone; and as soon you cannot see yourself any more in the mirror, withdraw, moving backwards, while saying: "Pater, in manus tuas commendo spiritum meum." [Father, into your hands I commend my spirit.] This bone you must keep.

Sometimes it was necessary to summon the help of demons, which was always a matter fraught with danger. A medieval manual of demonic magic tells the magician to go to a field and inscribe a circle on the ground, fumigate it and sprinkle it, and himself, with holy water while reciting Psalm 51:7 ('Cleanse me with hyssop, and I shall be clean . . .'). He then conjures several demons and commands them in God's name to do his bidding by bringing him a cap of invisibility. One of them will fetch this item and exchange it for a white robe. If the magician does not return to the same place in three days, retrieve his robe and burn it, he will drop dead within a week. In other words, this sort of invisibility was both heretical and hazardous. That is perhaps why instructions for invisibility in an otherwise somewhat quotidian fifteenth-century book of household management from Wolfsthurn Castle in the Tyrol have been mutilated by a censorious reader.

Demons are, after all, what you might expect to find in a magical grimoire. The *Grimorium Verum (True Grimoire)* is another eighteenth-century fake attributed to Alibeck the Egyptian; it was alternatively called the *Secret of Secrets*, an all-purpose title alluding to an encyclo-paedic Arabic treatise popular in the Middle Ages. 'Secrets' of course hints alluringly at forbidden lore, although in fact the word was often also used simply to refer to any specialized knowledge or skill, not necessarily something intended to be kept hidden. This grimoire says that invisibility can be achieved simply by reciting a Latin prayer – largely just a list of the names of demons whose help is being invoked, and a good indication as to why magic spells came to be regarded as a string of nonsense words:

> Athal, Bathel, Nothe, Jhoram, Asey, Cleyungit, Gabellin, Semeney, Mencheno, Bal, Labenenten, Nero, Meclap, Helateroy, Palcin, Timgimiel, Plegas, Peneme, Fruora, Hean, Ha, Ararna, Avira, Ayla, Seye, Peremies, Seney, Levesso, Huay, Baruchalù, Acuth, Tural, Buchard, Caratim, per misericordiam abibit ergo mortale perficiat qua hoc opus ut invisibiliter ire possim . . .

. . . and so on. The prescription continues in a rather freewheeling manner, suggesting that one might, if so inclined, include a conjura-tion using characters written in bat's blood, before calling on yet more demonic 'masters of invisibility' to 'perform this work as you all know how, that this experiment may make me invisible in such wise that no one may see me'.

A magic book was scarcely complete without a spell of invisibility. One of the most notorious grimoires of the Middle Ages, called the *Picatrix* and based on a tenth-century Arabic work, gives the following recipe.* You take a rabbit on the '24th night of the Arabian month', behead it facing the moon, call upon the 'angelic spirit' Salmaquil, and then mix the blood of the rabbit with its bile. (Bury the body well – if it is exposed to sunlight, the spirit of the Moon will kill you.) To make yourself invisible, anoint your face with this blood and bile at night-time, and 'you will make yourself totally hidden from the sight of others, and in this way you will be able to achieve whatever you desire'.

'Whatever you desire' was probably something bad, because that was usually the way with invisibility. A popular trick in the eighteenth century, known as the Hand of Glory, involved obtaining (don't ask how) the hand of an executed criminal and preserving it chemically, then setting light to a finger or inserting a burning candle between the fingers. With this talisman you could enter a building unseen and take what you liked, either because you are invisible or because everyone inside is put to sleep.

These recipes seem to demand a tiresome attention to materials and details. But really, as attested in *The Book of Abramelin* (said to be a system of magic that the Egyptian mage Abramelin taught to a German Jew in the fifteenth century), it was quite simple to make yourself invisible. You need only write down a 'magic square' – a small grid in which numbers (or in Abramelin's case, twelve symbols representing demons) form particular patterns – and place it under your cap. Other grimoires made the trick sound equally straightforward, albeit messy: one should carry the heart of a bat, a black hen, or a frog under the right arm.

Perhaps most evocative of all were accounts of how to make a ring of invisibility, popularly called a Ring of Gyges. The twentieth-century French historian Emile Grillot de Givry explained in his anthology of occult lore how this might be accomplished:

The ring must be made of fixed mercury; it must be set with a little stone to be found in a lapwing's nest, and round the stone must be

---

* Appearing hard on the heels of an unrelated discussion of the Chaldean city of Adocentyn, it betrays the cut-and-paste nature of many such compendia.

engraved the words, "Jésus passant ✠ par le milieu d'eux ✠ s'en allait."
You must put the ring on your finger, and if you look at yourself in a
mirror and cannot see the ring it is a sure sign that it has been successfully manufactured.

Fixed mercury is an ill-defined alchemical material in which the liquid
metal is rendered solid by mixing it with other substances. It might
refer to the chemical reaction of mercury with sulphur to make the
blackish-red sulphide, for example, or the formation of an amalgam
of mercury with gold. The biblical reference is to the alleged invisibility
of Christ mentioned in Luke 4:30 ('Jesus passed through the midst of
them') and John 8:59 (see page 155). And the lapwing's stone is a kind
of mineral – of which, more below. Invisibility is switched on or off
at will by rotating the ring so that this stone sits facing outward or
inward (towards the palm), just as Gyges rotated the collet.

Several other recipes in magical texts repeat the advice to check in
a mirror that the magic has worked. That way, one could avoid embarrassment of the kind suffered by a Spaniard who, in 1582, decided to
use invisibility magic in his attempt to assassinate the Prince of Orange.
Since his spells could not make clothes invisible, he had to strip naked,
in which state he arrived at the palace and strolled casually through
the gates, unaware that he was perfectly visible to the guards. They
followed the outlandish intruder until the purpose of his mission
became plain, whereupon they seized him and flogged him.

Some prescriptions combined the alchemical preparation of rings
with a necromantic invocation of spirits. One, appearing in an eighteenth-century French manuscript, explains how, if the name of the
demon Tonucho is written on parchment and placed beneath a yellow
stone set into a gold band while reciting an appropriate incantation, the
demon is trapped in the ring and can be impelled to do one's bidding.

Other recipes seem to refer to different *qualities* of invisibility. One
might be unable to see an object not because it has vanished as though
perfectly transparent, but because it lies hidden by darkness or mist,
so that the 'cloaking' is apparent but what it cloaks is obscured. Or
one might be dazzled by a play of light (see page 25), or experience
some other confusion of the senses. There is no single view of what
invisibility consists of, or where it resides. These ambiguities recur
throughout the history of the invisible.

Partly for this reason, it might seem hard to discern any pattern in these prescriptions – any common themes or ingredients that might provide a clue to their real meaning. Some of them sound like the cartoon sorcery of wizards stirring bubbling cauldrons. Others are satanic, or else high-minded and allegorical, or merely deluded or fraudulent. They mix pious dedications to God with blasphemous entreaties to uncouthly named demons. That diversity is precisely what makes the tradition of magic so difficult to grasp: one is constantly wondering if it is a serious intellectual enterprise, a smoke-screen for charlatans, or the credulous superstition of folk belief. The truth is that magic in the Western world was all of these things and for that very reason has been able to permeate culture at so many different levels and to leave traces in the most unlikely of places: in theoretical physics and pulp novels, the cults of modern mystics and the glamorous veils of cinema. The ever-present theme of invisibility allows us to follow these currents from their source.

Demons did it

Magic was first an adjective. It was the quality attributed to those whom the Greeks called *magoi*, who came from Persia and the East with mysterious, exotic skills that evoked both wonder and apprehension. 'Those itinerants', says social historian Barbara Maria Stafford, 'specialized in accessing the invisible'. But what became known as magic drew on other resources too: it was a crossing point for science and religion, for intellectual and popular culture and also for Jewish, Muslim, Eastern, Christian and pagan beliefs. What might not emerge from such a rich brew?

Yet until the thirteenth century, the consensus was that there was only one way to do magic and that was to enlist the help of demons. Catholics and Protestants in the early Reformation would not admit to agreeing on much, but they were united in the long-standing view that demons existed: immaterial, invisible creatures, made of some kind of incorruptible quintessence, they wrought mischief and wickedness in the world. When storms gathered, demons filled the air and bells were rung to deter them. When illness, famine or disaster struck, unseen demons were responsible. They were masters of illusion, as the fifteenth-century French theologian Peter Mamoris attested:

From vapours and fumes demons can simulate bodies, they can effect figures and colours, they can divert the species of objects in the air so that they do not reach the eye and the object remains invisible.

In his authoritative manual *On the Intellect and Demons* (1492), the Italian Agostino Nifo says explicitly and on holy authority that demons can make a man invisible.

Their omnipresence explained all manner of illicit happenings. Demons were sexually voracious: they came unseen at night as incubi and succubae to copulate with men and women, stealing the male seed and placing it in the female womb without the knowledge of the individuals concerned. That is how Merlin was said to have been conceived, making him half-spirit. Witches mate with invisible devils, according to the infamous witch-hunter's manual *Malleus Maleficarum* (*Hammer of the Witches*) (1486) by the inquisitors Heinrich Kramer and Jakob Sprenger:*

> Manie times witches are seene in the fields, and woods, prostituting themselves uncovered and naked up to the navill, wagging and mooving their members in everie part, according to the disposition of one being about that act of concupiscence, and yet nothing seene of the beholders upon hir; saving that after such a convenient time as is required about such a peece of worke, a blacke vapor of the length and bignesse of a man, hath beene seene as it were to depart from hir, and to ascend from that place.

That these lustful demons were invisible could be convenient, however. In *The Discoverie of Witchcraft* (1584), the Englishman Reginald Scot remains deadpan and lets the facts tell the tale of an incubus's visitation:

> You shall read in the legend, how in the night time *Incubus* came to a ladies bed side, and made hot love unto hir: whereat she being offended, cried out so loud, that companie came and found him under hir bed in the likenesse of the holie bishop *Sylvanus*, which holie man was

---

* Sprenger's authorship is contested; some historians say his name was added purely to lend gravitas to the volume, Sprenger being the august Dean of Theology at the University of Cologne.

much defamed thereby, untill at the length this infamie was purged b
the confession of a divell made at S. *Jeroms* toombe.

One assumes it was the bishop himself who secured this confession.

Only natural

As the Renaissance dawned, there arose a new way to understand
magic: that, while it drew on unseen influences, these were not necessarily demonic. In the tradition known as natural magic, nature itself
was infused with invisible, *occult* forces that caused marvellous effects.
These forces rationalized a whole suite of 'philosophical arts' that
today seem to exemplify the credulousness of that age: alchemy,
astrology, divination. But the aims of natural magic were primarily
practical, even mundane: it was a system by means of which useful
matters could be accomplished, whether making metals and medicines
through alchemy, or constructing ingenious machines, or hiding things
from sight. According to Pico della Mirandola, the precocious fifteenth-
century Italian scholar who typified the spirit of Renaissance humanism,
natural magic was 'the practical art of natural knowledge'.

There was no denying that occult forces – in the literal sense of
influences that are invisible or hidden – really do exist in nature. The
authoritative thirteenth-century theologians Albertus Magnus, William
of Auvergne and Thomas Aquinas all believed that the stars exerted
an occult influence on worldly affairs. Medieval writers debated
whether these astral forces defined our destiny or could be resisted.
But for such writers, 'magic' as such was still wicked deception at
best, and demonic meddling at worst.

In contrast, the defenders of natural magic such as Pico and his
mentor Marsilio Ficino, physician to Cosimo de' Medici in Florence,
insisted that mastery of the occult framework of natural magic was
nothing more than a question of acquiring a deep understanding of
nature: the objective today claimed by science. Yet natural magicians
were constantly accused of witchcraft. Because natural magic drew
on a long tradition of practical experimentation, any use of apparatus
and instruments was liable to arouse suspicions of necromancy. Roger
Bacon and Robert Grosseteste, two of the foremost practical investigators of natural phenomena in the thirteenth century, both suffered

from such dark rumours, notwithstanding that Bacon (known as 'Doctor Mirabilis') was a devout Franciscan friar and Grosseteste was the bishop of Lincoln. Three centuries later Reginald Scot took pains to make the distinction clear:

> In this art of naturall magicke, God almightie hath hidden manie secret mysteries; as wherein a man may learne the properties, qualities, and knowledge of all nature. For it teacheth to accomplish maters in such sort and oportunitie, as the common people thinketh the same to be miraculous; and to be compassed none other waie, but onlie by witch-craft. And yet in truth, natural magicke is nothing lese, but the worke of nature. For in tillage, as nature produceth corne and hearbs; so art, being natures minister, prepareth it.

That was the natural magician's credo: we are doing nothing that nature cannot do, but are merely assisting it on its way. Just as metals were thought to mature naturally in the earth into silver and gold, so the alchemist reproduced this natural process in his alembics and retorts.

Scot was one of the bold individuals who, in an age when witches were being persecuted and burnt, insisted that they were nothing but deluded fools who had no genuine communion with devils at all. It was from Scot that Aubrey got the account of Jewish merchants attempting an invisibility spell; by relating its ridiculous failure, Scot was discrediting the whole idea of demon-juggling sorcery.

Perhaps the best attested occult force was magnetism. What could seem more magical than a compass needle seeking north, or a lode-stone drawing iron to it without making contact? Magnetism was a marvel that spawned many legends. The king of Egypt Ptolemy Philadelphus in the third century BC was said to have sought the help of the Macedonian architect Dinocrates to suspend an iron statue of his wife Arsinoë* so that it floated in the air within her tomb, using a roof made of magnets. Both Ptolemy and Dinocrates died before the task was completed, but there are echoes of this tale in the apoc-ryphal legend that the coffin of Mohammed was levitated magnetically.

---

* She was also Ptolemy's sister; so it went in ancient Egypt.

The sixteenth-century Italian philosopher and doyen of natural magic Giambattista della Porta recounted that idea in his book *Magiae Naturalis* (1558), and claimed to have made such an arrangement himself using an iron object constrained by a thread from below, whereupon 'it will hang in the air and tremble and sway itself'. There are tales of magnetic islands that pull ships towards them; to pass by, boats would have to be made with wooden nails, since iron ones would be pulled out and the vessel would collapse.

The word magnet derives from the region of Magnesia on the Aegean Sea, where lodestone can be found, but it might also share an etymological root with magic itself. In the Middle Ages the Latin word for diamond, *adamas*, came to also be used for magnets, and is said to be linked to the French *aimant*, love – for the attraction of iron and magnet was commonly viewed as a kind of love, or as natural magicians would put it, sympathy. That anthropomorphism of occult forces had a long tradition; in the fourth century the Alexandrian poet Claudian wrote that

> Iron and the lodestone are drawn together and united. What can be this subtil flame which, entering these two metals, can give rise to this sympathy?

Electrical attraction, seen in the way rubbed amber (Greek *elektron*) picks up small particles, was similarly magical. For the ancient Ionian philosopher Thales, the attractive powers of the magnet and of amber were evidence that these substances possessed a soul. Thomas Aquinas felt that the magnet's unseen mode of action rendered it beyond human comprehension. But later authors attributed these effects to some invisible fluid or 'effluvium' that passes through the air and draws other objects hither. 'It is probable', wrote William Gilbert, physician to Elizabeth I, in his groundbreaking study of magnetism *De magnete* (1600), 'that amber does exhale something peculiar to itself which allures bodies themselves'. The Anglo-Irish scientist Robert Boyle developed this idea in the seventeenth century. Some people, Boyle says, maintain that amber emits streams of fluid in which little vortices push lighter bodies towards the amber's surface. Others suppose that certain invisible rays issuing from the material cool and shrink, pulling small items back with them. This was starting to sound like science, yet it borrows ideas directly from the natural-magic tradition.

What natural magic sought to accommodate, then, was the undeniable fact that many, perhaps most, things that happen in the world have unseen causes. When you kick a ball, the motive force is clear: the foot pushes against the ball.* But whence the power that turns seeds into trees and fruits? What stirs the winds, what propels the stars and the wandering planets? In the scheme of natural magic, the universe is a network of occult forces, effluvia and emanations. This subtle web of connections was discerned by the third-century Greek philosopher Plotinus, the founder of the tradition based on Plato's cosmology and known as Neoplatonism. In *Enneads*, Plotinus wrote that

> All things must be enchained; and the sympathy and correspondence obtaining in any one closely knit organism must exist, first, and most intensely, in the All.

In other words, the Neoplatonic magical universe is organized according to principles of correspondence, for example so that certain plants and metals are associated with particular planets that govern their behaviour in an invisible conspiracy of sympathies. As the famed German magician and physician Heinrich Cornelius Agrippa explains in his *Occult Philosophy* (1531–33),

> There is therefore a wonderfull vertue, and operation in every Hearb and Stone, but greater in a Star, beyond which, even from the governing Intelligencies every thing receiveth, and obtains many things for it self.

These relationships were often revealed in the outward forms of nature: they could be read literally from the surface of things. The sun's powers ruled all things golden and ruddy: gold itself, amber, honey and cinnamon. Plotinus explained that

* Actually even this is not so obvious. Yes, the foot moves the ball while the two are in contact, but what invisible agency keeps the ball moving once they have become separated? And why does the ball eventually cease moving apparently of its own accord? Medieval philosophers believed that during contact the foot imparts to the ball a quantity called impetus, which, like a kind of fuel, is slowly consumed.

All teems with symbol; the wise man is the man who in any one thing can read another, a process familiar to all of us in not a few examples of everyday experience.

There are antipathies too: after all, magnets will attract iron but not copper, and (at least according to Plutarch) amber will attract all light bodies except basil.* The key point was that, although these powers are occult, the natural magician can discern them, master them, and use them. 'Behold the herbs!', wrote Agrippa's contemporary, the Swiss alchemist Paracelsus, who drew heavily on Neoplatonic ideas. 'Their virtues are invisible and yet they can be detected . . . nothing is so hidden in them that man cannot learn it.'

It was a system that worked by analogy: just as this, so that. Just as the magnet loves iron, so the earth attracts the falling stone. (Gravity was widely thought to be a kind of magnetism; Isaac Newton assumed as much in his early work on the subject.) These analogies linked events in the macrocosm of the world to processes in the human body or the alchemist's flasks. Rain falls just as steam condenses in a vessel (which is true) and rivers adopt the same branching channels as blood flowing in the body (which is also basically true). Such analogies were generally qualitative – no one worried, for example, that the force of magnetism seemed to be so much stronger than the force of gravity, so that a lodestone could pick up an iron nail in defiance of the gravitational tug of the entire earth. In this much, natural magic affirms sociologist Theodor Adorno's notion of the occult as 'the readiness to relate the unrelated'. All the same, its invisible forces offered a vision of a law-bound universe in which specific effects have identifiable and natural causes.

The *Picatrix*, that ill-reputed medieval compendium of magic, explained how to capture and channel the influences of the planets to work marvels – some of them impressive, such as filling a room with snakes, others much more banal, such as curing toothache. This may be accomplished, the book said, by harnessing the forces that are 'hidden from the senses, so that most people do not grasp how they

---

* Specific exceptions to the electrostatic tug of amber were commonly asserted. The Greek physician Ctesias in the fifth century BC wrote of trees in India that would attract everything *except* amber; one sometimes came across sheep stuck to their roots.

happen or from what causes they arise'. Some of the things that practitioners of natural magic believed might strike us as absurd today, but the basis of that belief was no longer miraculous. A natural magician might induce a rain of frogs, but not because he conjured frogs out of nothing; they were ordinary frogs, borne aloft by an invisible force. A necromancer, in contrast, would have had demons put them up there.

Natural magic thereby provided a rational basis for astrology, alchemy, meteorology, biology and a great deal else. By controlling and manipulating this web of attractions and repulsions, there seemed no limit to what the magician might achieve. Concentrating and directing the virtues of herbs and minerals, he could make medicines. Using lenses and mirrors, he could produce spectacular displays of light. He could create life, fly, conjure up voices and apparitions. And all this was done not by compelling invisible demons but by learning how to direct the tendencies of nature.

Because natural magic drew on the kind of practical skills and beliefs that had long existed in folk tradition, it was sometimes a strange mix, for popular beliefs were, as they always have been, a blend of sound empirical facts and wild superstitions. At first natural magicians had no systematic way of distinguishing one from the other, and were apt to end up dignifying the most bizarre ideas within the framework of occult forces and sympathies. It was hard to decide what or who to trust – the world was strange enough to make almost anything seem plausible. Some magical practices and prescriptions combine a seemingly naturalistic manipulation of herbs or minerals with the kind of incantation that, in folk belief, would have been thought to enhance the potency of the object – itself a legacy of the religious power of prayer and ritual.

Natural magic, then, had a strong pragmatic aspect; it was in some ways a form of experimental science. Although a tradition of philosophical 'experiment' is often traced back to Roger Bacon in the thirteenth century, manual manipulations of matter and mechanism had always been practised by craftspeople and were refined in the Arabic alchemy and medicine of the seventh to the tenth centuries. The magnetic experiments of William Gilbert led him to conclude that the earth is itself a giant magnet that aligns the poles of the compass needle. He pointed out that the attractive power of amber was shared by many other substances, including glass, sulphur, sealing wax, resins, rock salt and alum. He tested the old belief that magnetism was nullified by smearing a lodestone with the juice of onions or garlic, and found that it wasn't so.

To test: that was something new. In the late sixteenth century, advocates of natural magic began to turn to genuine experiment, attempting to verify or discredit the popular claims made for occult powers. If it was not yet science's famous (if sometimes exaggerated) reliance on falsification tests of theories, and if there was at first little concern with quantitative measurement, all the same this was a less credulous approach than that of the traditionalists who merely regurgitated supposed facts on the authoritative testimony of Aristotle or Pliny. Magic was less inclined to defer to the questionable wisdom of antiquity. Early scientists like Galileo, Boyle and Newton followed the example, borrowing and modifying the scheme of invisible forces as they did so.

Making magic

Many of the recipes for invisibility from the early Renaissance onward therefore betray an ambiguous credo. They are often odd, sometimes ridiculous, and yet there are indications that they are not mere mumbo-jumbo dreamed up by lunatics or charlatans, but hint at a possible rationale within the system of natural magic.

It's no surprise, for example, that eyes feature prominently among the ingredients. From a modern perspective the association might seem facile: you grind up an eyeball and therefore people can't see you. But to an adept of natural magic there would have been a sound causative principle at work, operating through the occult network of correspondences: an eye for an eye, you might say. A medieval collection of Greek magical works from the fourth century AD known as the *Cyranides* contains some particularly grotesque recipes of this sort for ointments of invisibility. One involves grinding together the fat or eye of an owl, a ball of beetle dung and perfumed olive oil, and then anointing the entire body while reciting a selection of unlikely names. Another uses instead 'the eye of an ape or of a man who had a violent death', along with roses and sesame oil. An eighteenth-century text spuriously associated with Albertus Magnus (he was a favourite source of magical lore even in his own times) instructs the magician to 'pierce the right eye of a bat, and carry it with you and you will be invisible'. One of the cruellest prescriptions instructs the magician to cut out the eyes of a live owl and bury them in a secret place.

A fifteenth-century Greek manuscript offers a more explicitly optical theme than Aubrey's head-grown beans, stipulating that fava beans are imbued with invisibility magic when placed in the eye sockets of a human skull. Even though one must again call upon a pantheon of fantastically named demons, the principle attested here has a more naturalistic flavour: 'As the eyes of the dead do not see the living, so these beans may also have the power of invisibility.'

Within the magic tradition of correspondences, certain plants and minerals were associated with invisibility. For example, the dust on brown patches of mature fern leaves was said to be a charm of invisibility because it was thought to carry the fern's invisible principle of reproduction: unlike other plants, they appeared to possess neither flowers nor seeds, but could nevertheless be found surrounded by their progeny.

The classical stone of invisibility was the heliotrope (sun-turner), also called bloodstone: a form of green or yellow quartz (chalcedony) flecked with streaks of a red mineral that is either iron oxide or red jasper. The name alludes to the stone's tendency to reflect and disperse light, itself a sign of special optical powers. In his *Natural History*, Pliny says that magicians assert that the heliotrope can make a person invisible, although he scoffs at the suggestion:

> In the use of this stone, also, we have a most glaring illustration of the impudent effrontery of the adepts in magic, for they say that, if it is combined with the plant heliotropium, and certain incantations are then repeated over it, it will render the person invisible who carries it about him.

The plant mentioned here, bearing the same name as the mineral, is a genus of the borage family, the flowers of which were thought to turn to face the sun. How a mineral is 'combined' with a plant isn't clear, but the real point is that the two substances are again bound by a system of occult correspondence.

Agrippa repeated Pliny's claim in the sixteenth century, minus the scepticism:

There is also another vertue of it [the bloodstone] more wonderfull, and that is upon the eyes of men, whose sight it doth so dim, and dazel, that it doth not suffer him that carries it to see it, & this it doth not do without the help of the Hearb of the same name, which also is called Heliotropium.

It is more explicit here that the magic works by dazzlement: the person wearing a heliotrope is 'invisible' because the light it reflects befuddles the senses. That is why kings wear bright jewels, explained Anselm Boetius, physician to the Holy Roman Emperor Rudolf II in 1609: they wish to mask their features in brilliance. This use of gems that sparkle, reflect and disperse light to confuse and blind the onlooker is attributed by Ben Jonson to the Rosicrucians, who were often popularly associated with magical powers of invisibility (see pages 32–3). In his poem *The Underwood*, Jonson writes of

> The Chimera of the Rosie-Crosse,
> Their signs, their seales, their hermetique rings;
> Their jemme of riches, and bright stone that brings
> Invisibilitie, and strength, and tongues.

The bishop Francis Godwin indicates in his fantastical fiction *The Man in the Moone* (1634), an early vision of space travel, that invisibility jewels were commonly deemed to exist, while implying that their corrupting temptations made them subject to divine prohibition. Godwin's space-voyaging hero Domingo Gonsales asks the inhabitants of the Moon

> whether they had not any kind of Jewell or other means to make a man invisible, which mee thought had beene a thing of great and extraordinary use . . . They answered that if it were a thing faisible, yet they assured themselves that God would not suffer it to be revealed to us creatures subject to so many imperfections, being a thing so apt to be abused to ill purposes.

Other dazzling gemstones were awarded the same 'virtue', chief among them the opal. This is a form of silica that refracts and reflects light to produce rainbow iridescence, indeed called opalescence.

Whether *opal* derives from the Greek *opollos*, 'seeing' – the root of 'optical' – is disputed, but opal's streaked appearance certainly resembles the iris of the eye, and it has long been associated with the evil eye. In the thirteenth-century *Book of Secrets*, yet again falsely attributed to Albertus Magnus, the mineral is given the Greek name for eye (*ophthalmos*) and is said to cause invisibility by bedazzlement:

> Take the stone *Ophthalmus*, and wrap it in the leaf of the Laurel, or Bay tree; and it is called *Lapis Obtalmicus*, whose colour is not named, for it is of many colours. And it is of such virtue, that it blindeth the sights of them that stand about. Constantius [probably Constantine the Great] carrying this in his hand, was made invisible by it.

It isn't hard to recognize this as a variant of Pliny's recipe, complete with cognate herb. In fact it isn't entirely clear that this Ophthalmus really is opal, since elsewhere in the *Book of Secrets* that mineral is called *Quiritia* and isn't associated with invisibility. This reflects the way that the book was, like so many medieval handbooks and encyclopedias, patched together from a variety of sources.

Remember the 'stone from the lapwing's nest' mentioned by Grillot de Givry? His source was probably an eighteenth-century text called the *Petit Albert* – a fabrication, with the grand full title of *Marvelous Secrets of Natural and Qabalistic Magic*, attributed to a 'Little Albert' and obviously trading once more on the authority of the 'Great Albert' (Magnus). The occult revivalist Arthur Waite gave the full account of this recipe from the *Petit Albert* in his *Book of Ceremonial Magic* (1913), which asserts that the bird plays a further role in the affair:

> Having placed the ring on a palette-shaped plate of fixed mercury, compose the perfume of mercury, and thrice expose the ring to the odour thereof; wrap it in a small piece of taffeta corresponding to the colour of the planet, carry it to the peewit's [lapwing's] nest from which the stone was obtained, let it remain there for nine days, and when removed, fumigate it precisely as before. Then preserve it most carefully in a small box, made also of fixed mercury, and use it when required.

Now we can get some notion of what natural magic had become by the time the *Petit Albert* was cobbled together. It *sounds* straightforward enough, but who is going to do all this? Where will you find the lapwing's nest with a stone in it in the first place? What is this mysterious 'perfume of mercury'? Will you take the ring back and put it in the nest for nine days and will it still be there later if you do? The spell has become so intricate, so obscure and vexing, that no one will try it. The same character is evident in a nineteenth-century Greek manuscript called the Bernardakean Magical Codex, in which Aubrey's instructions for growing beans with a severed head are elaborated beyond all hope of success: you need to bury a black cat's head under an ant hill, water it with human blood brought every day for forty days from a barber (those were the days when barbers still doubled as blood-letters), and check to see if one of the beans has the power of invisibility by looking into a new mirror in which no one has previously looked (does such a thing exist?). If the spell doesn't work (and the need to check each bean shows that this is always a possibility), it isn't because the magic is ineffectual but because you must have done something wrong somewhere along the way. In which case, will you find another black cat and begin over? Unlikely; instead, aspiring magicians would buy these books of 'secrets', study their prescriptions and incantations and thereby become an adept in a magical circle: someone who possesses powerful secrets, but does not, perhaps, place much store in actually putting them to use. Magical books thus acquired the same talismanic function as a great deal of the academic literature today: to be read, learnt, cited, but never used.

Matter of gravity

There was nothing about occult forces in themselves that conflicted with the prevailing Aristotelian view of natural philosophy in the Renaissance. Aristotle drew distinctions between the substance of bodies and their qualities, and those qualities could be either manifest (obvious to the onlooker) or occult (things that couldn't be deduced directly). As the Aristotelian Dutch philosopher Franco Burgersdijk put it, matter has

> quality . . . manifest or occult: the former affects the senses in itself; the latter is perceived only from effects, and sympathies and antipathies are to be referred to it.

For example, the medical properties of particular herbs were occult because, while you could see their effects (in principle), you couldn't deduce their presence by inspecting the herb itself.

The French philosopher Marin Mersenne, a hard-headed rationalist and colleague of Descartes,* felt that to say that a substance possessed occult powers was simply to say that one didn't know the reasons behind them. 'These qualities', he said, 'are occult only to the ignorant, for learned people . . . do not use these terms, showing that what one calls occult is evident to them; and if there are qualities that they do not know, they freely admit their ignorance.' All the talk of sympathies and antipathies by the occult philosophers, in contrast, is just wordplay 'to cover their defeats'.

For adherents of the so-called mechanical philosophy like Mersenne and Descartes, invisible forces such as magnetism and electricity must have some explanation that involved the interactions of invisibly small particles and mechanisms. 'There are no qualities which are so occult,' wrote Descartes, 'no effects of sympathy or antipathy so marvellous or strange, nor any other thing so rare in nature . . . that its reason cannot be given.' No longer should it be necessary to accept that occult forces and qualities were 'just how things are'.

The Cartesian French philosopher Pierre Gassendi clarified what these mechanisms must consist of: 'tiny invisible instruments . . . to do the job of pulling or pushing . . . tiny hooks, strings, goads and poles . . . which, even though invisible and impalpable, are not undescribable.' The mechanical philosophy sought to replace the unknown with the familiar: away with spirits and emanations and in their place, particles rubbing and bumping, or effluvia like liquids that were ultimately atomistic, or tiny devices that operated like keys, locks, pins, hooks and mills. Those mechanisms were already understood, and could be trusted.

The problem was that this picture was no less speculative than that of the occultists. The mechanical philosophers were doing just what Mersenne claimed they were not: asserting unproved mechanisms rather than admitting ignorance. When Gottfried Leibniz accused the

---

* Even the most rationalistic souls were products of their age. Mersenne believed in demonic possession and in various magical phenomena, saying for example that he knew of 'a gentleman . . . who thickens the air so much that he can walk on it.'

English occult philosopher Robert Fludd, the *bête noire* of the mecha-
nists, of 'fabricat[ing] faculties or occult qualities . . . like little demons
or imps [to] perform whatever is wanted', he failed to recognize that
the active atoms and atom-sized machines of Descartes were scarcely
any different. Just like occult forces, the mechanical philosophy could
explain anything, offering spurious endorsement of effects that were
fantastical to begin with. Some claimed that the reason we could not
see these 'atomical effluviums' for ourselves was that our sight had
been corrupted by the Fall: for prelapsarian Adam, the causes of invisible forces had been visually evident.

It was in the context of these tensions that Isaac Newton sought
to explain gravity as an occult force. Or rather, he argued merely that
a force exists which holds the planets in place, although we know
nothing of its cause. He called it a 'power', which might be carried
by some non-mechanical fluid in the cosmos (an 'ether'), or might
simply be 'seated in the frame of nature by the will of God'. This
image seemed congruent with the claim of the Platonists such as
Henry More who held court at Newton's university in Cambridge:
that the world is pervaded by 'a substance incorporeal', a kind of
pan-animistic world spirit. It also sounded dangerously close to Fludd's
spiritus mundi, which he made the ultimate source of life and identified as the essence of Christ. In any event, for Newton, gravity might
reveal divine action in the world. That was his way: searching for God
in nature, looking for nature in the word of God.

There was nothing particularly heterodox in this vision of God
acting through a beneficent, invisible force. It was a commonplace of
seventeenth-century theology that God exercised providential and
active control over events on earth. That was the true provenance of
Adam Smith's famous Invisible Hand that purportedly maintains
economic stability: as historian Peter Harrison has said, 'almost
certainly, when readers encountered the phrase in Smith, they would
have understood it as referring to God's unseen agency in political
economy'– whether Smith intended it or not.* Humans, like planets,
were deemed to be led by God's invisible hand to accomplish His ends.

---

* It seems appropriate that the neoliberal conviction in the ability of the unchecked
market to bring about economic stability turns out to have its roots in an expression
of religious faith.

Newton's rival Leibniz accused him of abandoning the mechanical precepts of Boyle and Descartes in favour of Fludd's mysticism: injecting occult principles into bodies. He wasn't being fair, although that was nothing unusual in this famously bitter rivalry. Newton had known that the vision of gravity offered in his *Principia* (1687) – a force acting at a distance without apparently being conveyed by direct contact between particles – would be controversial. Yet what distinguished his invisible force from that of Neoplatonists like Fludd – and this was precisely how the new experimental science in general diverged from natural magic – was not so much a question of whether occult influences existed, nor even how they operated, but whether they were autonomous phenomena or symbols of something else. While natural magic was a metaphorical system relying on analogy and correspondence, Newton's gravity represented nothing but itself. What's more, he was here speaking of a 'force' in much the same way as modern physicists do: as a handy concept for talking about effects that can be captured by mathematical equations. The precise nature of the force was a question one could postpone until later; what mattered, what was 'real', was the quantitative law that described it. 'These Principles', he wrote, referring to gravity 'I consider not as occult Qualities . . . but general Laws of Nature'. Talk of forces of attraction and repulsion, Newton knew, 'will displease many' because of their association with the occult philosophies. But he insisted that he was using such terms merely as a way of speaking about visible effects, not as an explanation of causes. He went so far as to say that genuine action-at-a-distance was impossible:

> 'Tis utterly inconceivable, that inanimate brute Matter (without the mediation of some Immaterial Being) should operate upon and affect other Matter without mutual Contact; that distant Bodies should act upon each other through a *Vacuum* without the intervention of something else by and through which the action may be conveyed from one to the other.*

* The same tensions persist today in explaining gravity. Albert Einstein's description of it, in the theory of general relativity, as the curvature of spacetime, accounts for how instantaneous gravitational action at a distance is possible. But it is also generally accepted that gravity can only be made consistent with quantum theory by invoking a hypothetical particle (the graviton) that mediates the interaction, just as

When it came to invisible forces, Leibniz could do no better than Newton anyway. His explanation for them invoked immaterial entities called monads, which infuse bodies without being exactly a part of them and which act like purposeful agencies, making the world a shadow play of tiny minds. Not only was it teeming with what were in effect little sentient creatures, but it might even consist of them: 'perhaps the block of marble itself', said Leibniz, 'is only a mass of an infinite number of living bodies like a lake full of fish.' He had taken Fludd's mystical 'world soul' and cut it up into little pieces. Even though Leibniz disputed the claims of natural magicians to be able to influence the play of these interactions, nonetheless historian Brian Copenhaver says that Leibniz, of all people, 'gave the occult philosophy a last hour of respectability'.

Were Leibniz's monads part of the physical or the spiritual world – atoms or angels, both invisible but in different ways? It was a question barely even formulated at that time and Leibniz felt no compunction to raise it. When John Locke did so in his *Essay Concerning Human Understanding* (1689), it was only in order to postpone it. We have no real understanding of either corporeal or spiritual substance, he said, beyond what we can say about their qualities: the former has weight and shape, say, the latter thinks and feels. Yet merely by making this distinction, Locke was already moving away from Leibniz's vision of animate corpuscles, cleaving spirit from flesh. When Kant wondered at the end of the eighteenth century if monads could be seen in the microscope, he was just being facetious.

So it seemed that, faced with the evident fact of occult interactions, no one could do without some sort of explanatory (but invisible) agency, whether it be magical sympathies and emanations, Cartesian vortices, the 'corpuscles' and 'effluvia' invoked by Robert Boyle, Newton's powers, Leibniz's monads, or the will of God. Although the mechanical philosophy's explanation of occult forces looks much more scientific and rational than old notions of correspondences and sympathies, it was every bit as reliant on hypothetical, unseen influences. In the nineteenth century the mechanism of autonomous force was reinstated by the Scottish physicist James Clerk Maxwell in the form

---

all the other fundamental forces have particulate forms: the photon of light, for example, is the intermediary of electromagnetic force.

of a *field* – now one of the central concepts of modern physics. These fields have tiny particles associated with them. (Much has recently been made of the discovery of corpuscles of the so-called Higgs field, which instils mass much as gravity instils weight.) But Maxwell's fields too retain a whiff of occult natural magic, as we shall see. Today we accept invisible emanations and forces without demur: they bind atoms and molecules, hold shut the refrigerator door and enable us to talk to one another from mountain-tops. And like natural magicians we can control and manipulate them, and work wonders.

Invisible brotherhoods

Because of its disreputable taint, not to mention its innate habits of secrecy and concealment, natural magic was often practised in private. In the late sixteenth century, like-minded individuals started to gather in 'closed' societies and academies to explore this approach to understanding nature. One of the most famous was the Accademia dei Lincei in Italy, a transitional organization between magic and science, which boasted Giambattista della Porta and Galileo among its members. These groups were often regarded as hotbeds of heresy, satanic arts and radical politics, and were outlawed by the church or the state. Yet some survived and evolved into scientific societies, elitist but visible in the public sphere and devoted to an experimental approach to natural philosophy. Others took a more mystical direction, propagating 'occult knowledge' with an explicitly magical (not to mention politically radical) orientation. The most notorious, and ultimately one of the most enduring, was the Rosicrucian movement that began in Germany in the early seventeenth century.

Rosicrucianism was launched in the German city of Cassel with the publication of a leaflet called the *Fama fraternitatis* (*Report of the Brotherhood*) in 1614, followed the next year by the *Confessio fraternitatis* (*Confession of the Brotherhood*). These documents, published anonymously but widely thought now to be the work of the Protestant alchemist Johann Valentin Andreae of Herrenberg, asserted that this mysterious brotherhood had been established in the fifteenth century by one Christian Rosencreutz, a Dutchman who had learnt magical secrets in the East. The manifestos called on the hidden members of the sect, allegedly existing throughout Europe, to come forth and spread this knowledge. In truth these documents sowed mostly confusion, rumour, deceit and unrest; for several years the brotherhood was imagined to be lurking everywhere, and in that fractious time, with the devastating Thirty Years' War about to unfold, no one could be sure what they were plotting.

That the Rosicrucian cult was hidden was a central aspect of its constitution. Not only did its members claim to possess esoteric and occult knowledge, but they passed unseen in society. No one was sure how to identify or contact the murky syndicate, and that naturally made people uneasy. When in 1623 the citizens of Paris found the city walls postered with the declaration that '[w]e, the deputies of the principal College of the brethren of the Rose-Cross have taken up our abode, visible and invisible, in this city', many were alarmed, even if some intellectuals dismissed the affair as a hoax. One anti-Rosicrucian pamphlet claimed to expose '[t]he frightful Compacts entered into between the Devil and the pretended "Invisibles"' – for that was how the Rosicrucians became known in France.

These shady figures were said to possess a genuine magic of invisibility, granted to them by the devil, enabling them to enter into any place no matter how securely bolted and fastened its portals. Some of the French declarations (there were several) stated as much explicitly; for example,

> All those who seek entrance into our society and congregation [will be transformed] from visible beings into invisible, and from invisible into visible, and they shall be transported into every foreign country to which their desire may lead them.

Another hostile pamphlet reported that the Rosicrucians 'assert that they can become invisible at pleasure, a quality incommunicable to any natural body which consists of matter and form, and one which can never be acquired by any legitimate science'. The writers of this document, evidently Catholics who suspected a Protestant plot, went on to say 'they frequent the Sabbaths, cherish toads, make poisonous powders, dance with fiends, raise tempests, ravage fields, destroy orchards, assassinate and torture their neighbours by the infliction of innumerable diseases'. Everything, in other words, traditionally associated with witches.

To avoid being branded a Rosicrucian, public figures needed to dispel any suggestion that they possessed such diabolical powers. Descartes was apparently keen to contact the Rosicrucians when he heard about them during his travels in Germany – but on returning to France and discovering how ill they were regarded there, he was said to have foregone his usual solitary habits and visited his friends often, lest he be thought to be invisible himself.

The motif of a hidden society of savants appealed to intellectuals like Descartes. In letters of 1646–47 Robert Boyle mentioned an Invisible College who 'do now and then honour me with their company'. No one knows who this mysterious group was, although a common and natural supposition is that they were the London-based gathering of experimental philosophers who formed the nucleus of the Royal Society. Another possibility is that they were the utopian philosophers centred around the Prussian Samuel Hartlib and the Bohemian Jan Amos Comenius, both exiles from the Thirty Years' War. The overlap in aims and imagery of the latter group with the Rosicrucians led the historian Frances Yates in the 1970s to propose a Rosicrucian origin not only for the Royal Society but for the Enlightenment in general, an idea now largely dismissed by contemporary historians. Others have suggested more fancifully that Boyle was referring to a lodge of Freemasons. Historian Charles Webster plausibly identifies the Invisible College with a group of chemically inclined Anglo-Irish intellectuals gathered around Boyle's sister Lady Ranelagh. But the fact is that the Invisible College still remains true to its name.

The Freemasons – an offshoot of the Rosicrucians, sharing their carefully preserved social invisibility – originated in the seventeenth century* and did overlap with the Royal Society: several of its first members, such as Robert Moray, Elias Ashmole and possibly Christopher Wren, were Freemasons. Freemasons too were apt to be

---

* Freemasonry asserts a connection to the lodges or professional gatherings of the medieval masons responsible for designing and building churches and cathedrals. While the origins of Freemasonry are unclear, the link with craftsmen who actually worked with stone is tenuous. Freemasonry in its modern form, clear records of which begin around the seventeenth century, was a secret society concerned with esoteric philosophy: this is called Speculative Freemasonry, distinguished from the Operative Freemasonry of the stone-workers.

associated with magical invisibility. In 1752, for example, one John Macky was examined by the Grand Committee of Freemasons in London for pretending 'to teach a Masonical Art, by which any man could (in a moment) render himself invisible.' It would have been all too easy to find an audience for such claims.

Hocus pocus

The historian of the occult – of natural magic, secret societies, hidden forces – is constantly confronted with the prospect of trickery: false potions and cures, hoax conspiracies, casually implausible claims. For magic has always had an ambivalent connection to showmanship, and teems with charlatans, imposters and prestidigitators. Some of this was innocent enough. Thomas Betson, a fifteenth-century monk of Syon Abbey in Middlesex, England, wrote down instructions for making images appear as if out of nowhere using elaborate arrangements of mirrors, along with accounts of how to make an egg jerk about apparently of its own accord (by attaching a fine hair to it) and an apple move in like fashion (by hollowing out its core and putting a beetle inside). These were guileless tricks – the means would doubtless be hidden from the audience, but there was no indication that Betson saw them as anything other than amusements (which makes one wonder if his clerical masters knew how he was passing his time).

The status of this so-called 'juggling' can be gauged from the magician's manual *Hocus Pocus Junior* (1634), which claimed to set forth 'the art of juggling . . . plainly and exactly, so that an ignorant person may thereby learn the full perfection of the same, after a little practise'. The book revealed the tricks of the eponymous performer, the stage name of one William Vincent, who in 1619 was given a licence to practise the 'art of Legerdemain' (sleight of hand) throughout England and Ireland. It contained such stalwarts as the 'three cup-and-ball trick', as well as instructions for 'how to make a stone seem to vanish out of your hand'. The cod Latin 'Hocus Pocus' possibly satirizes the way that the Catholic mass functioned for many worshippers as a kind of ill-understood magical incantation (although the claim of the Archbishop of Canterbury John Tillotson in 1695 that it is a corruption of *hoc est corpus meum* seems unlikely). Jugglers like Vincent would eat glass, stick objects into their flesh, produce nails, daggers

The title page of *Hocus Pocus Junior*, a seventeenth-century manual of stage magic.

and fire from their mouths and perform the celebrated decapitation illusion on their accomplices. Each trick would be accompanied by a little Latin incantation, the better to suggest arcane powers.

Illusions in which objects were moved by what appeared to be invisible forces – what would later be called telekinesis – were particularly popular. There are medieval accounts of loaves dancing and fish jumping out of pans. How audiences perceived these performances is not clear. Did they collude with a suspension of disbelief for the sake of entertainment, or did they truly think they were witnessing marvels? It's not clear that such distinctions were in fact recognized at all. On the one hand, magic blended seamlessly with mechanical wizardry. On the other hand, attributing genuinely magical powers to a prestidigitator was not then so obviously different from the way we accept today that stage magicians have special skills and knowledge that we do not. When Chaucer's Franklin in the *Canterbury Tales* tells how a squire tries to court the wife of another man by acquiring the services of a natural magician to make the rocks of Brittany's coast vanish (she having rashly promised him her love if he could accomplish this seemingly impossible feat), he draws no distinction between the deceitful or whimsical illusions of 'jugglers' (from the Latin *joculare*, to jest, and meaning anyone who performs wondrous tricks) and genuine magical disappearance. 'I am certain', the squire's brother muses,

> that there must be sciences
> By which illusions can be made, appliances
> Such as these subtle jugglers use in play
> At banquets.

When Chaucer's wizard conducts his (successful) 'experiment' of invisibility on the rocks – 'by illusion or apparition – call it jugglery' – he prepares for it with astrological arithmetic, implying that he is indeed manipulating the occult forces of nature to spirit away the rocks. This is at the same time the duplicitous trickery of the juggler and the mysterious natural magic of the adept. Even in the seventeenth century the hidden mechanisms of astrological influences were not distinguished from the concealed hydraulics, strings and levers of the mechanical marvels then in vogue among the aristocracy of Europe – witness the book on ingenious automata published in 1648 by John Wilkins, a founder of the Royal Society, which was called *Mathematical Magick*. This elision is not surprising, for often both operations were effected by the same people: the astrologer and mathematician John Dee combined a serious interest in occult forces with the preparation of elaborate illusionistic apparatus for the Elizabethan stage.

Thus Reginald Scot's claim that jugglers 'always acknowledge wherein the art consisteth' might be a little optimistic. These mountebanks would not want to be suspected of witchcraft – but as magicians have always known, magic always needs a little mystery. Its practitioners, whether they were itinerant swindlers or philosophical magi like Dee, rarely made clear distinctions between preternatural power, mechanics and deception. As mechanical wizardry became a part of the Renaissance court philosopher's skill, it was recommended that they leave the machinery hidden, the better to enhance the wonder of the performance.

By the eighteenth century there was a well developed aesthetic of what Barbara Maria Stafford calls 'the visible invisible': illusionistic trickery that used visible effects to imply invisible causes. Now the rationalistic stage magician vied with the charlatan in exploiting a public craving for spectacle: 'Both the instructor and the mountebank manipulated gadgetry to visibilize an invisible realm,' Stafford says.

Even if, therefore, Daniel Defoe felt that his *Compleat System of Magick: or, the History of the Black-Art* (1736) was documenting a moribund intellectual tradition, the truth was that magic was instead making the transition from an occult science to a form of stagecraft. Following the lead of *Hocus Pocus Junior*, magicians advertised themselves openly as prestidigitators who achieve their illusions by manual and mental manipulation rather than marvellous or forbidden knowledge. They toured fairs, theatres and clubs, and began to forgo the head-chopping and body-piercing for less gruesome and more wondrous spectacles, now performed in everyday clothes rather than the mountebank's motley. The occult web of natural magic was unweaving; the world was becoming disenchanted.

Magic revealed

In place of natural magic there was now a new kind of science, grounded in experiment and generally explained in terms of mechanical theories of particles in contact with one another. In the late eighteenth century, electrical phenomena entered the experimental scientist's repertoire, and the magician-performers saw at once the possibilities they offered. The Frenchman Nicolas-Philippe Ledru, who adopted the stage name Comus after the Greek god of anarchy and revelry, toured the courts of Europe entertaining nobles with his tricks and experiments involving electricity, magnetism and light. Ledru also styled himself a public educator, explaining the scientific principles behind his demonstrations and thus bridging the tradition of legerdemain and the emerging art of scientific spectacle. These stage magicians now openly acknowledged that what they did was mere trickery, and audiences came hoping to be astonished and even a little frightened but no longer fearing they were witnessing a black art. Manuals such as Henry Breslaw's *Magic Companion* (1784) were the successors of *Hocus Pocus Junior*, and if not all these books were as forthcoming as

Breslaw's, that was simply because of the value of preserving trade secrets. Philip Astley's *Natural Magic, or Physical Amusements Revealed* (1785) does not always fulfil the title's promise – Astley had, after all, to maintain the pole position of the Circus of England, the first of the modern circuses, of which he was the founder – but it leaves no doubt that sleight of hand is all that is involved in such trickery.

The impresario Étienne Gaspard Robert, a native of Liège in modern-day Belgium, who toured under the name Robertson, was another of these demystifiers. A playbill for one of his shows in 1802 advertised his intention 'to expose the Practices of artful Impostors and pretended Exorcists, and to open the Eyes of those who still foster an absurd Belief in ghosts or disembodied spirits'.* Similarly, in 1805 William Pinchbeck, a descendant of the renowned English maker of clocks and automata Christopher Pinchbeck, published *The Expositor; or, Many Mysteries Unravelled*, a manual of stage magic meant to amuse and 'to convince superstition of her many ridiculous errors'. Ventriloquism, a staple of such shows, was revealed as a common means by which fraudsters summoned invisible presences. One conjuror explained in 1800 how he would create 'spirits' by burning a mixture of antimony, sulphur and other chemicals and throwing his voice to make these fiery creations speak. The eminent scientist David Brewster showed what 'natural magic' had come to imply with his *Letters on Natural Magic* (1832), an explanation of hallucinations and illusions created by mirrors, ventriloquism, magic lanterns and other optical effects.

This was the golden age of theatrical magic and it flourished first in France. In the mid-nineteenth century Jean-Eugène Robert-Houdin performed at the Palais-Royal in Paris, where his 'experiments in natural magic' were said to be based on sound scientific principles. Several of Robert-Houdin's illusions were copied by the German performer Compars (known as Carl) Herrmann, who set up his own show in the 1840s in which he billed himself as the 'First Professor of Magic'. Compars trained his much younger brother Alexander and they toured together until Compars retired in the 1870s. Alexander brought the

---

* All the same, business is business, and Robertson, like several of his contemporaries, trod a fine line. According to Barbara Maria Stafford, he 'resembled the charlatans Franz Anton Mesmer [see page 42] and Alessandro conte di Cagliostro . . . in carefully keeping secret the means by which his illusions were contrived'.

show to the famous theatre of magic at the Egyptian Hall in London, where he was succeeded in his short tenancy by the Englishman John Nevil Maskelyne, a watchmaker who begat a celebrated dynasty of British stage magicians (and who was, incidentally, the inventor of the pay toilet). In 1905 Maskelyne and a group of other British magicians founded the Magic Circle, dedicated to the art of stage magic and illusion. Maskelyne was also a debunker of frauds: he exposed the American spiritualists Ira and William Davenport, explained the Indian rope trick and attacked Madame Blavatsky's brand of modern mysticism in *The Fraud of Modern 'Theosophy' Exposed* (1912).

In his introduction to Albert Allis Hopkins's now classic manual *Magic: Stage Illusions and Scientific Diversions, Including Trick Photography* (1898), the American writer and amateur magician Henry Ridgely Evans proclaimed that 'Science has laughed away sorcery, witchcraft, and necromancy'. Hopkins himself takes care not to antagonize Spiritualists, but he was clearly highly sceptical of their alleged occult powers, and they would have to be pretty dull-witted to miss his irony. One of Hopkins's informants (a Mr Caulk) carries out the 'rope test', in which the hands of the medium are bound in order to make sure that they cannot manipulate devices, and in so doing Caulk declares that he has never found one who was not an imposter. To this, the deadpan Hopkins suggests that 'this may be due to the bad fortune of the writer'.

Hopkins shows how stage magicians of the Victorian era made avid use of the newest scientific discoveries. Wilhelm Röntgen's X-rays, he noted, discovered barely three years before the book was published, 'are now competing with the most noted mediums in the domain of the marvellous'. These rays will cause light to be emitted from all manner of materials: porcelain, glass, enamels and any objects coated with standard phosphors such as zinc sulphide. Hopkins describes a trick in which a man dining alone is suddenly cast into darkness, whereupon he vanishes (except for his glowing eye-glasses) and the audience sees, seated across the table, a 'sinister guest': a shining skeleton, lit up by a hidden X-ray generator. Hopkins is evidently having fun when he proclaims that '[w]e have, therefore, only the trouble of selection in order to get up a "spirit séance" with every certainty of success, while genuine spiritual séances fail in most cases, as is well known, because the spirits are in an ill mood and disposed to be coyish'.

A ghoulish visitor at the dinner table made visible using X-rays in the dark, from Albert Hopkins's *Magic: Stage Illusions and Scientific Diversions, Including Trick Photography* (1898).

The comeback of the occult

It might seem odd that, after the demise of natural magic in the seventeenth century, the likes of Maskelyne were still finding it necessary to expose claims of occult powers in the early twentieth. But the fact is that the occult tradition itself did not wither during the Enlightenment: it merely took new forms, and has continued to do so ever since.

In 1895, after a century of scientific advance that began with the groundbreaking discoveries of Antoine Lavoisier and Michael Faraday and now stood poised for the revolution of quantum physics, William Butler Yeats felt able to write this:

I cannot get it out of my mind that this age of criticism is about to pass, and an age of imagination, of emotions, of moods, of revelation, about to come in its place; for certainly belief in a supersensual world is at hand.

Yeats' remark appeared in an essay titled 'The Body of the Father Christian Rosencrux'; he was one of the new occultists of the *fin de siècle*. This is one of the seeming paradoxes of modernity: that the blossoming of science coincided with a resurgence of interest in spiritual matters, mysticism and magic and new notions of invisibility. That this is no coincidence at all is a matter explored in Chapter 4.

Belief in invisible forces had survived and even flourished during the Enlightenment, thanks in considerable measure to the activities of the German physician (and Freemason) Franz Anton Mesmer, who asserted that human bodies contained an unseen magnetic fluid that could be channelled to influence other bodies. Mesmer began by using actual magnets in his medical treatments, but soon claimed that he could induce the same effects by merely making stroking motions over the patient's body, which harnessed a pervasive, cosmic force called animal magnetism. Mesmer was seemingly a master hypnotist who could induce trances ('magnetic sleep') or convulsions in his patients, giving rise to the term mesmerism. In the 1770s he developed a variety of bizarre practices and devices, most notably the *baquet*: a wooden tub covered with a lid and allegedly charged with the curative magnetic fluid, which would be conveyed to the afflicted limbs and organs of patients when they placed

A man presses a leg to the *baquet* to receive Mesmer's 'animal magnetism' treatment.

The sole remaining *baquet* apparatus can be seen at the Musée d'Histoire de la Médecine et de la Pharmacie in Lyon.

these parts against iron rods protruding from the lid, all in a séance-like atmosphere of dim light, tinkling music and incense.

Accused of fraud – and perhaps sexual misconduct – in Vienna after having claimed to cure a blind female musician, Mesmer fled the city and set up a practice in Paris. In 1784 Louis XVI appointed two commissions to investigate animal magnetism that included Lavoisier and Benjamin Franklin. They concluded that the phenomenon was illusory, after which it became the subject of satires and plays in France. But mesmerism was too entrenched in public culture by that stage for its popularity to suffer very much from these expert judgements.

Another invisible agency, called the odylic or odic force, was invoked in the mid-nineteenth century by the Prussian chemist Carl von Reichenbach, who claimed that it could be discerned by especially sensitive people as a kind of flame or aura visible around humans, magnets, crystals and plants. Reichenbach suspected that this force might be responsible for apparent sightings of ghosts, and reported that one young woman who he accompanied at night into a graveyard saw 'a delicate, fiery, as it were, a breathing flame' on a grave. The writer Edward Bulwer-Lytton drew on these speculations about an all-encompassing force, somewhere between electromagnetism and psychic energy, for his novel *The Coming Race* (1871), which describes a subterranean society called the Vril-ya who have harnessed a power called *vril*. As the narrator explains,

I should call [*vril*] electricity, except that it comprehends in its manifold branches other forces of nature, to which, in our scientific nomenclature, differing names are assigned, such as magnetism, galvanism, etc. These people consider that in *vril* they have arrived at the unity in natural electromagnetic forces.

This fictional *vril*, simultaneously a unified field theory and a magical force of the will through which all things are possible, has been even more fertile of legend than the supposedly real animal magnetism. Conspiracy theories insist that a Vril Society flourished in Germany as a precursor to the Nazi Party and that this secret society managed to tap into *vril* to power flying saucers and other secret weapons deployed by the Nazis, even allowing them to create an underground base on the moon. *Vril* finds a rather more prosaic incarnation (literally) in the name of the meat extract Bovril, where it was meant to suggest that this substance could be energizing. The fact that the Canadian company behind Bovril could draw on the association when it relaunched its 'liquid beef' product in 1886 shows how deeply Bulwer-Lytton's concept had penetrated into popular culture.

Although they were always controversial and sometimes dismissed as illusions of the imagination, animal magnetism and the odic force remained in linguistic currency in science until the late nineteenth century. When he used an electromagnet to make a treasure chest miraculously heavy, Robert-Houdin claimed the trick was done by 'mesmeric power'. Once the subtle, unseen alleged influences of the paranormal are given names – especially if they are linked to undeniable forces such as magnetism and gravity and framed within the scientific lexicon of the times – it has always proved tremendously hard to dislodge them. You can destroy reputations and throw labels into disrepute, but the basic idea is sure to emerge again, whether it is as orgone energy or morphic resonance. Our appetite for invisible agencies to explain events at the threshold of detectability or plausibility is, it seems, far greater than our desire to test the credibility of the events in the first place. When the English clergyman John Webster wrote in *The Displaying of Supposed Witchcraft* (1677) that 'men ought to be cautious and to be fully assured of the truth of the effect before they adventure to explicate the cause', he was speaking to every age.

However bogus their claims, Reichenbach and Mesmer aimed to

position themselves within the scientific mainstream. The occult tradition, however, also retained a more Gnostic aspect in which adepts possessed special, revelatory knowledge, exemplified by the early seventeenth-century Lutheran mystic Jacob Boehme and perpetuated during the following century in the figures of Claude-Louis, Comte de Saint-Germain, a would-be alchemist who traded his chemical expertise at the courts of Europe, and the wily Count Alessandro di Cagliostro, the assumed name of Italian apothecary Giuseppe Balsamo. A Freemason and trickster, Cagliostro was regarded with such suspicion that he became accused of involvement in the notorious Affair of the Diamond Necklace in which Marie Antoinette was accused of trying to defraud the crown jewellers, fuelling the discontent that led to the French Revolution. It seems the grounds for including Cagliostro in the conspiracy were nothing more than that he had been seen wearing diamonds around Paris and, after spending many months in the Bastille, he was acquitted. But his Freemasonry alone was enough to get him arrested later in Rome by the Inquisition and he spent his final years in prison.

The revival of the esoteric tradition in the nineteenth century arguably began with the publication of a book of natural magic called *The Magus, or Celestial Intelligencer* (1801), written by self-styled Rosicrucian and 'professor of chemistry' Francis Barrett, which was essentially a compilation gleaned from the writings of Renaissance magi such as John Dee, Agrippa and della Porta. The resurgence gathered pace in the mid century, thanks in large part to the efforts of the Frenchman Alphonse Louis Constant, who wrote under a Hebrew approximation to his name as Eliphas Lévi. His books *Dogme de la Haute Magie* (1854) and *Ritual de la Haute Magie* (1856) provided a compendium of occult lore remodelled for modern sensibilities. Concealment and invisibility were central to this doctrine.

As a young man Lévi planned to enter the priesthood, but he abandoned his training, developed into a political radical and was briefly imprisoned for the provocative pamphlets he published in the revolutionary year of 1848. Lévi's mysticism was a volatile blend of Christianity, Jewish Kabbalah, eastern esoterism, Freemasonry, Rosicrucianism, alchemy, mesmerism and Spiritualism, tailored to accommodate anyone sympathetic to such ideas. With his robes and full beard, he looked like a mage to some, and the devil to others.

Lévi conceived of an all-pervasive invisible fluid called astral light,

Eliphas Lévi (1810–1875), nineteenth-century master of the occult.

which he linked with electromagnetism, Mesmer's animal magnetism and Reichenbach's odic force. He called astral light 'a natural and divine agent, at once corporeal and spiritual, an universal plastic mediator, a common receptacle for vibrations of movement and images of form, a fluid and a force which may be called, in a sense at least, the Imagination of Nature'. This light, he said, 'warms, illuminates, magnetises, attracts, repels, vivifies, destroys, coagulates, separates, breaks and conjoins every-thing' – which, as is typical of these putative invisible forces, made it capable of explaining just about anything.

In the late nineteenth century, beliefs in occult powers blended with an interest in paranormal and Spiritualist phenomena. The unseen agents that some held responsible for phenomena such as table-rapping, telekinesis and spirit messages were variously identified as the immaterial essences of the dead, the astral bodies of the living, or a race of preternatural, non-human creatures. Theosophy, a mystical movement initiated in the United States by the Ukrainian émigré Helena Petrovna Blavatsky, asserted that there exists an astral plane

Helena Blavatsky (1831–1891), creator of Theosophy.

positively teeming with spirit forms: human souls, both of the living and the dead, in various stages of spiritual advancement or degradation, as well as nature spirits, the spirits of animals, higher non-human beings called Devas (which in former times were deemed to be angels) and artificial elementals such as demons made by black magic. Where Lévi commanded attention through the sheer bulk of his physical presence, Blavatsky seems to have possessed an almost hypnotic personality: her stare mesmerizes even in photographs. She launched her cult in 1875, claiming to be receiving messages from Spirit Masters who were the wise ancestors of the human race. One of her first converts and staunchest supporters was Henry Steel Olcott, a former colonel of the American Civil War and member of the committee that investigated Lincoln's assassination.

Blavatsky and Olcott moved to India in 1879, where they infused their quasi-religion with strains of Hinduism and Buddhism, before settling in London in 1887. In India they met the English writer Alfred Percy Sinnett, who returned to England to become head of the London Lodge of Theosophists. The former priest Charles Leadbeater also became influential in the London group; he travelled to Southeast Asia in the mid-1880s, touring Burma and Ceylon with Olcott. Blavatsky laid out her esoteric philosophy in *The Secret Doctrine* (1888), a manifesto that fuses religion, science and philosophy and provides a template for the New Age tracts of modern times. For Blavatsky, just about every force, entity and existence worth considering in the universe is invisible. As with Lévi's brand of magical mysticism, there was nothing the Theosophical doctrine could not accommodate, from gravity and atomic theory to gods, angels, submerged continents and Buddhist karma.

The new occultism revitalized Rosicrucianism and Freemasonry. All of these hidden societies attracted a particular type of person: romantic, idealistic, intense, suspicious of positivistic science and eager to find more in the world than meets the eye (and, one has to say, all too often devoid of the slightest hint of humour or self-mockery). Such personalities tend to be individualists, rarely capable of sustaining a stable community and apt to inflate small disagreements into major doctrinal conflicts. As a result, these occult organizations were fissiparous, spawning a bewildering array of seemingly minor variations through a toxic mixture of academic spat and religious crusade. The social reformer Annie Besant, for example (see page 122), joined the London Theosophists before setting up her own rival branch, founded the first Grand Lodge of Freemasonry that accepted both men and women (Mixed Freemasonry), and then created the Order of the Temple of the Rose Cross as a short-lived offshoot of the Grand Lodge.

Samuel Liddell Mathers represents another example of this brand of irrepressible egotism. He was just the kind of eccentric that such movements accrue, reinventing himself by adding MacGregor to his surname in a probably spurious attempt to claim Scottish heritage (he was born in Hackney, London) and gravitating towards every esoteric organization on offer. Married to the sister of the anti-rationalist philosopher Henri Bergson, Mathers was initiated into the Freemasons in 1877 before becoming a member of the Rosicrucian Society. He then went on to create his own mystical organization, the Hermetic Order of the Golden Dawn, in 1888 in collaboration with fellow Freemasons William Robert Woodman and William Wynn Westcott (also a Theosophist). Members of the Golden Dawn included the scholar of mysticism Arthur Edward Waites, the ill-famed magical adept Aleister Crowley, W. B. Yeats and the children's writer E. Nesbit (Edith Bland). They claimed to possess paranormal powers acquired in ceremonial rituals – including a ritual of invisibility, in which the magician would surround himself with a shroud of shadow by evoking and banishing spiritual beings, accompanied by exhortations along the lines of 'Come to me, O shroud of darkness and night'.

The self-importance and petulant in-fighting of the Golden Dawn makes it hard not to find much of the business of this latter-day occultism amusing and tiresome in equal measure. Old photos of Mathers got up in what are alleged to be ancient Egyptian robes don't inspire confidence in his claims to have rediscovered ancient wisdom. Yet the Golden Dawn and other occult revivalists unearthed and translated a considerable body of historically important literature and helped to make occultism a subject of serious scholarship. Mathers, for example, produced translations of *The Sacred Magic of Abramelin the Mage* and other manuals of magic, while Waite waded patiently through the copious and verbose texts of Paracelsus. On the whole they believed in something akin to natural magic, but in a formulation reconfigured for the age of Jung and Freud: Besant called it 'the use of the Will to guide the powers of external nature'.

During the birth of modernism, the occult revival touched every sphere of art. It infused the Symbolist movement in visual art and poetry. Even though Virginia Woolf lampooned Annie Besant in *The Waves* (1931) – she was unimpressed by a lecture Besant delivered in London in 1917 – and was rather contemptuous of mystics generally, the poet and literary scholar Julie Kane asserts that if one were to catalogue the various types of mystical experience appearing in Woolf's writings, 'the list would be virtually indistinguishable from the topics of interest to the Theosophists and spiritualists of her day: telepathy, auras, astral travel, synesthesia, reincarnation, the immortality of the soul, and the existence of a Universal Mind'.

For visual artists, occult mysticism seemed to supply a sympathetic language for their exploration of subjective experience. In his influential tract *Concerning the Spiritual in Art* (1912), Wassily Kandinsky called Theosophy 'one of the greatest spiritual movements' towards a union of Eastern and Western mysticism. His concept of the artist as a pianist who, touching the keys of form and colour, elicits a vibration in the human soul has a clear affinity with Blavatsky's theories, while his discussions of colour (deeply indebted to Goethe's idiosyncratic ideas) are saturated with Neoplatonic notions of correspondence and sympathy. (These are doubtless linked to Kandinsky's synaesthesia, which made him see colours in response to non-visual stimuli such as music.)

The Dadaists and the Surrealists explored magical modes of thinking, which Max Ernst called 'the means of approaching the unknown by other ways than those of science and religion'. Ernst read Agrippa ('a splendid magician') and in an autobiographical sketch he spoke of his 'excursions in the world of marvels, chimeras, phantoms, poets, monsters [and] magi'. Max Ernst, this self-portrait declared, 'died the 1st of August 1914. He resuscitated the 11th of November 1918 [the dates of the First World War] as a young man aspiring to become a magician and to find the myth of his time'. For the founder of the Surrealists, André Breton (who read Eliphas Lévi closely), poetry was a magic art, a view made explicit in his book *L'Art Magique* (1957). The Surrealists, mindful of atomic theory and Einstein's new conception of time and space, felt that in any case science and the occult were converging. The French writer Pierre Mabille claimed in 1940 that while modern physicists 'use the most precise methods and possess the most powerful instruments . . . they are nonetheless the legitimate heirs to the tradition of the marvellous' – the tradition of the Renaissance magus. That was not a one-sided view; the physicist Arthur Eddington, whose observations of a solar eclipse in 1919 verified Einstein's theory of general relativity, was insistent on affinities between mysticism and science. And surely much of that perception of shared ground came from a common interest in what could not be seen. In 1949, Gabrielle Buffet-Picabia, the wife of one-time Surrealist Francis Picabia, proclaimed, with some justification, that every field of intellectual inquiry was then attempting to capture the 'non-perceptible'.

Modern magic

Invisibility and disappearances are stock features of modern-day illusionists' performances, from David Copperfield to David Blaine. The thrill comes now not from witnessing miraculous occult powers but from the dexterity and daring of what we know to be trickery. Yet it's a common mistake to imagine that this artifice is all that remains of the legacy of natural magic.

For magic is not so much a technical skill as a mode of thinking. Anthropologists have long regarded it this way, and early pioneers of that discipline helped to establish magic as a genuine cultural phenomenon rather than a consequence of individuals' ignorance and credulity. This doesn't mean, however, that they were able to avoid belittling and marginalizing it. Interpreting cultures faddishly in an evolutionary (and inevitably racist) framework, late-nineteenth-century anthropologists such as Edward Burnett Tylor and James Frazer presented magic as a pre-scientific mode of thought among 'savages'. To Tylor it was 'one of the most pernicious delusions that ever vexed mankind'; to Frazer, the 'bastard sister of science'.

But gradually magic came to be seen as fulfilling beneficial roles too – which were by no means limited to 'primitives'. Emile Durkheim pointed out that magic still persists in science-based societies, whether (in today's terms) as 'irrational' beliefs in UFOs and crystal healing or as forms of magical thinking that we engage in every day: if I follow this routine, I will be protected from illness. Magic is not a procedure that produces effects, but a symbolic system with a social function.

The Polish-British anthropologist Bronislaw Malinowski recognized this: in perhaps the most fertile formulation of magic, he explained that it 'raises the psychological self-confidence of its believers', letting them dare to believe they can control events that might otherwise seem ineffable and inevitable, and thereby establishing a base on which to construct a genuinely efficacious technology. Magic, he said, fills the gap left by the absence of science, and in this way it 'ritualizes man's optimism' – it is 'the embodiment of the sublime folly of hope' – the hope that, one way or another, *this can be done.*

Can invisibility magic be done? Magic supplies the dream – but what happens when technology tries to impinge on realms hitherto occupied by myth? There is, as we shall see, already ample experience to show us what to expect.

3 Fear of Obscurity

Of all the *Arcana* of the invisible World, I know no One Thing
about which more has been said, and less understood, than this
of Apparition: It is divided so much between the Appearance of
good and the Apparition of bad Spirits, that our Thoughts are
strangely confus'd about it.

Daniel Defoe
The Secrets of the Invisible World Disclos'd (1729)

Insensibly, we yielded to the occult force that swayed us, and
indulged in gloomy speculation. We had talked some time upon
the proneness of the human mind to mysticism, and the almost
universal love of the Terrible, when Hammond suddenly said to
me, 'What do you consider to be the greatest element of Terror?'

Fitz-James O'Brien
'What Was It?' (1859)

The actor David Garrick had a set-piece during his performances of
Hamlet, the role for which he was most famous, that electrified London
theatre audiences in the eighteenth century. It came when the Prince
of Denmark sees his father's ghost at the start of the play: 'Look my
lord, it comes.' The *St. James Chronicle* said in 1772 that 'As no Writer
in any Age *penned* a Ghost like Shakespeare, so, in our Time, no Actor
ever *saw* a Ghost like Garrick'. After witnessing a performance, the
German scientist and aphorist Georg Christoph Lichtenberg wrote
that 'His whole demeanour is so expressive of terror that it made my
flesh creep even before he began to speak'. Samuel Johnson quipped

that such a reaction – head jerked back, eyes staring wildly – was enough to frighten a ghost itself.

Garrick is shown in the midst of this tour-de-force in a painting of 1756 by the artist (and, intriguingly, electrical scientist) Benjamin Wilson – which is now lost, but is reproduced in a contemporaneous print. Doesn't it seem here as if Garrick's hair is actually rising from his scalp? And so it should, for that is exactly what happened. But making one's hair stand on end at will was beyond the capability of even this ingenious performer. Garrick achieved the spine-tingling effect (which goes by the splendid name of horripilation) with the aid of a London wig-maker named Perkins, who created a mechanical wig powered by hydraulics – a contrivance that embarrassed some of Garrick's biographers, who felt betrayed by such gimmickry in a supposedly serious actor.

But although Garrick was not, by today's standards, above a certain degree of hamminess, the wig was not just a cheap trick. Garrick is credited with supplanting the mannered, overblown acting style that

David Garrick as Hamlet, on seeing his father's ghost in Act I. Mezzotint after a painting by Benjamin Wilson, 1756.

prevailed at the time by a realistic mode which convinced the audience that the characters he portrayed had a genuine emotional world. It seems likely that this naturalistic approach was informed by Garrick's Cartesian view of human physiology, in which the body was regarded as a kind of hydraulic mechanism driven by fluids called animal spirits that were pumped around the organs and limbs. Within this view, an artificial hydraulic wig was little different from the way real horripilation was thought to work by a rush of fluids to the head. Like all emotion, it was simply a matter of biomechanics, an orchestration of gestures and grimaces. That conception would explain the complaint of one of Garrick's fiercest critics, the actor and playwright Theophilus Cibber, who jeered at 'his over-fondness for extravagant attitudes, frequently affected starts, convulsive twitchings, jerkings of the body, sprawling of the fingers, flapping the breast and pockets [and] his pantomimical manner'.*

Yet there is another defence of Garrick's now seemingly risible 'fright wig': he needed all the help he could get, for he had set himself the task of conjuring the illusion of the ghost by gesture alone. Whereas previously the dead king was generally played by an actor, Garrick insisted that the ghost of Elsinore should be invisible: a disembodied voice whose presence was seen only by the actors. But theatrical invisibility is a difficult trick – as film makers later discovered, it needs visible signifiers of an unseen presence to sustain the illusion.

Garrick's choice represented a decision not just about staging but about what the ghost in *Hamlet* – on which of course the whole plot turns – truly means. It is a statement about how the play should be interpreted. For we are then forced to ask: is this a real spirit, or just a figment of Hamlet's tortured mind?

Shakespeare's ghosts

Ghosts were a common, even clichéd sight on the Elizabethan stage. They served as narrators, popping up at opportune moments to fill in a bit of back-story. As such, they were no cause for alarm either for

---

* Cibber was not an unbiased witness: he was the son of the playwright and actor Colley Cibber, whose florid performances belonged to the tradition that Garrick's new style was threatening to eclipse.

their implication or their appearance. They were represented by a sort of Jack-in-the-box puppet, or else by an actor with whitened face, dressed in warm clothes made of furs. These theatrical narrator-ghosts were a device borrowed from the plays of the Roman dramatist and statesman Seneca, which supplied a model for the revival of tragedy during the Renaissance. The Senecan ghost typically appeared in the prologue, calling for an act of revenge that motivated the play's tragic plot.

This function sounds much the same as that of the shade of Hamlet's dead father, and indeed Seneca's influence on Shakespeare has been well documented. But the ghost in Hamlet is no glove puppet. He is made to sound hardly less terrible to the audience than he is to Hamlet and his friends: the sight 'harrows me with fear and wonder', gasps Horatio. That's what Shakespeare did to the theatrical ghost: he made it real, humanized, haunting and disquieting. His spirits are truly spooky. According to the scholar of Elizabethan drama John Dover Wilson, 'Shakespeare managed . . . to lift the whole ghost-business on to a higher level'. The Ghost, he says, is 'the linchpin of *Hamlet*; remove it and the play falls to pieces.' Wilson's opinion is shared by scholars today. 'The ghost in *Hamlet* is like none other,' says literary professor Stephen Greenblatt, 'not only in Shakespeare but in any literary or historical text that I have ever read.'

No one needed to ask what manner of apparition the Senecan ghost is; it is, in Wilson's words, merely a 'bit of dramatic machinery'. But ghosts in Shakespeare, and in some of the Jacobean plays that came after, leave the audience guessing. Indeed, they leave the characters guessing. The ambiguity about his father's ghost is central to Hamlet's hesitation. Marcellus believes it to be a demon; Horatio doubts it exists at all. 'There is a whole genre of Elizabethan and Jacobean plays', says historian Keith Thomas, 'in which ghostly apparitions make their appearance, leaving us to speculate whether they are not really demons in human form rather than the human souls they purport to be.' This uncertainty about the ghost's status raises questions about its visibility: should we *see* the ghost or not?

The obvious way to answer that question is with another: what did Shakespeare himself intend? There has been a vigorous and colourful dispute about that in the rarefied world of Shakespeare studies. In the nineteenth century it was popular to suggest that his phantoms are subjective: they are seen only by those who need to see them, implying that they are psychological projections and not real spirits. In the early twentieth century there was a revolt against this idea. In 1907 the American Shakespeare scholar Elmer Edgar Stoll insisted dogmatically that Shakespeare's ghosts were the spectres of 'popular superstition', which pretty much everyone in Elizabethan times considered to be perfectly 'real', if intangible. But any distinction between what was 'real' and what was 'imagined' is itself something of an anachronism, given that imagination was invested with actual instrumental power – the ability to make things manifest – in the late sixteenth century. This is precisely what Theseus speaks of in *A Midsummer Night's Dream*: 'as imagination bodies forth / The forms of things unknown.'

Besides, what *was* the ghost of 'popular superstition' in Shakespeare's time? Although the answer is not simple, we can at least say that it was determined largely by your religion. Catholics believed that the souls of the dead reside for a time in Purgatory before being admitted (if they warrant it) to heaven. This gave souls a period in which to haunt the living: to be, as the king's ghost in *Hamlet* says, 'doomed for a certain time to walk the night'. But Protestants rejected the idea of Purgatory – which raises the question of how a dead soul can feature in what is undoubtedly a Protestant play. Might, then, the ghost be a demon masquerading as the king, to provoke Hamlet into acts of slaughter and, indirectly, Ophelia into sinful suicide?

This was the choice, it seems: ghosts were either dead souls, or they were demons – or maybe angels. All were real entities; as the Shakespeare scholar Robert Hunter West has said, when *Macbeth* was first performed in 1607 'Englishmen were seriously aware in a way that we are not of an invisible world about them'. Cornelius Agrippa assigned a trio of spirits (or perhaps a tripartite individual) to every person: a guardian angel responsible for salvation of the soul, a 'demon of geniture' who acts as custodian of astrological fortune and a 'demon of profession' who appears once a man has chosen his professional role. A man could learn to command these personal angels/demons, for example so that they would perform feats of illusion or conceal things by thickening the air. But ghosts, said Agrippa, were something else: the tormented vestiges of people who had led a wicked life, wandering the earth because they have either escaped from hell or been denied heaven. There was good reason to fear them, for they could possess the living.

*

In the late sixteenth century ghosts became a popular topic for writers, and several learned treatises on the topic were published at that time: *Of Ghosts and Spirits Walking by Night* (1572) by the Swiss Protestant theologian Ludwig (Lewes) Lavater, *A Treatise of Ghosts* (1588) by the Capuchin monk Noel Taillepied, and Pierre Le Loyer's *A Treatise of Spectres or Strange Sights, Visions and Apparitions Appearing Sensibly Unto Men* (1605). Even James I of England touched on the subject in his condemnation of occult arts *Daemonologie* (1597). Like the hardline Lavater, James dismissed the papist notion that ghosts are the souls of the dead released from Purgatory, asserting that they are all demons (Lavater at least admitted angels too).

Doctrinal point-scoring aside, for the most part these books were attempts at 'natural histories' in the manner that was popular at the time: efforts to provide a taxonomy of phenomena, along with naturalistic explanations for them. Taillepied insisted from the outset that we should 'distinguish calmly between natural happenings [such as noises at night] and ghosts'. 'Melancholicks and those suffering from the Hyp', he warns, 'can imagine all sorts of Visions which are often mere Fancy and Unreal', while 'Timorous and fearful Men may easily persuade themselves and imagine they can see and Hear most alarming Things, whereas in fact there is Nothing at all to Fear.' Ghosts might also be imputed to defective vision, natural entities such as glow worms, and men disguising themselves 'to Impose on others'.

But make no mistake: some ghosts are real, even if we might not always be able to see them. 'Sometimes', Taillepied wrote, 'a ghost will appear in a house and be only visible to the dogs, who slink and shiver, and immediately run to their masters for protection.' Alternatively, he says, 'the clear unclouded vision of a child often perceives spiritual visitants whom older eyes cannot discern.' He attests to the phenomenon now called poltergeists:*

Very often the servants in a house have at night heard unquiet Spirits, who seem to be moving the kitchen utensils and furniture . . . until one would verily believe that the whole place was turned topsy-turvy

---

* The first known use of the term poltergeist (German for 'noisy ghost') to describe an invisible ghost that causes a commotion is by Martin Luther in a pamphlet of 1530, where he attributes them to the deceptions of the Roman Church.

and everything spanwhengled and smashed to smithereens. In the morning all has been in apple-pie order, and not a kilterment displaced or disturbed.

Whether or not to remain invisible is a choice a ghost may make for itself, for St Augustine tells us (as Taillepied puts it) that spirits 'only clothe themselves with a body when they wish to appear, and they wear it as a garment'. What are these bodies made of? Condensed air, says the author, perhaps mingled with vapour and mist. The poltergeist slams doors, makes heavy footfalls, throws objects around and sometimes pinches and bruises people (although sexual assault by ghosts is a modern phenomenon). Its manifestation could also be detected by the stench of brimstone or decay, or by a sudden drop in temperature, or the fact that flames turn blue in its presence. All this was known to Shakespeare: 'The lights burn blue', says Richard III after Buckingham's ghost appears to him. 'It is now dead midnight. Cold fearful drops stand on my trembling flesh.'

The souls of the departed, according to Taillepied, may be sent back to earth by God to deliver a message. Pierre Le Loyer agreed, saying that apparitions 'do never show themselves but that they presage and fore-show something'. Shakespearian ghosts do always have motives and messages to impart, which perhaps only the intended recipients can perceive. 'This spirit, dumb to us, will speak to [Hamlet]', says Horatio. The notion of a ghost who, like Banquo in *Macbeth*, haunts the guilty party alone was well established. In 1654, for example, a privateer named John Baldock was driven to confess that he had robbed and killed an English soldier in Guernsey after being visited by the dead man's ghost, which no other person could see. We might be inclined now to attribute this to the fevered imaginings of a guilty conscience, but Shakespeare wasn't in contrast blindly literal about the selective visibility of spectres, for the powers of invocation and agency attributed to the imagination in the late Renaissance leave no clear distinction between a ghost being a projection of the mind and an objective (albeit supernatural) phenomenon.

The crucial issue, says Taillepied, is to distinguish between the benevolent messenger of God and the demon bent on wickedness. That, surely, is Hamlet's dilemma: should he trust the ghost or not? And this was not, then, just a quandary for the narrative, but impinged on the vigorous, politically charged debate about what ghosts are. For his part, Taillepied provides many tips on how to tell the difference between good and bad apparitions, some of which (is the ghost trying to sell you something?) remain worth heeding today for assessments of character.

Ghosts of the Enlightenment

Since scepticism about the existence of ghosts was already being expressed in the late seventeenth century, we might expect the Enlightenment to have been rather dismissive of spirits and spooks. But on the contrary, the eighteenth century saw ghosts awarded a new respectability, exemplified by Daniel Defoe's earnest treatise *The Secrets of the Invisible World* (1727), a Taillepied-style guide to ghosts and how to identify them. 'Between our Ancestors laying too much stress upon them', Defoe complained, 'and the present Age endeavouring wholly to explode and despise them, the World seems hardly ever to have come at a right Understanding about them.' If you fear that every apparition is the Devil, he says, how are you going to know him when he really *does* appear?

Besides, ghosts still had a social role to fulfil. As Keith Thomas explains, they 'personified men's hopes and fears, making explicit a great deal which could not be said directly.' They sanctioned proper conduct by ensuring that the dead were revered and by threatening horrible retribution for our sins. They are the invisible police, all-seeing agents that patrol norms and boundaries. In this respect traditional ghosts are not emissaries of chaos, but are on the contrary social conservatives. Isn't this just what Marley's ghost is up to, showing Scrooge the errors of his ways?

Yet Enlightenment rationalism could hardly fail to diminish credulity about popular ghost stories – or rather (for many intellectuals of that age felt the privilege of enlightenment was theirs alone) to foster scorn for the foolish beliefs of the rabble. Nothing better illustrated that foolishness than the notorious scandal of the Cock Lane Ghost of London. This arose from a dispute between one William Kent of Norfolk and his landlord, the parish clerk Richard Parsons. After the death of Kent's wife Elizabeth in childbirth, his short-lived baby son had been looked after by Elizabeth's sister Frances (Fanny), who then became romantically (and illicitly) involved with Kent. They moved to

London, where in 1759 Parsons, who many deemed to be a disreputable fellow and a drunkard, offered the couple lodgings in his home in Cock Lane, near St Paul's Cathedral. It was then, with Fanny now pregnant, that scratching and rapping noises began to be heard at the house.

Fanny died of smallpox before the child was born, and Kent moved out; by 1761 he was married again. But he had made Parsons a loan that the landlord had never repaid, and in early 1762 Kent finally recovered the debt after taking legal proceedings. Later that year, the mysterious noises at Cock Lane started again. Parsons, along with a cleric named John Moore from the nearby St Sepulchre's church, decided that, while the earlier disturbances must have been caused by Elizabeth's unquiet ghost, they were now being produced by Fanny's spirit. Communicating with this phantom through the classic 'one knock for yes' method, they 'discovered' that Fanny had in fact been poisoned with arsenic administered by Kent, who had inherited her estate. Hearing of these accusations, Kent returned to clear his name and a séance was convened to interrogate the ghost in Kent's presence. To an audience of several other witnesses, 'Scratching Fanny' repeated the charge of

A caricature of the Cock Lane Ghost from 1762 by an unknown artist, titled 'English Credulity, or the Invisible Ghost'.

poisoning and predicted that Kent would be hanged. 'Thou art a lying spirit, thou are not the ghost of my Fanny!' he exclaimed.

As word got out, the public became captivated by the affair. It transpired that Scratching Fanny made her rapping and scratching only when Parsons' young daughter was there; she was also called Elizabeth (Betty) and was prone to fits. The girl seemed to be the medium through which the ghost elected to communicate. As news spread, Parsons began to admit visitors to the Cock Lane residence (for an entrance fee) to 'talk' to Scratching Fanny via his daughter. Waiting crowds, some probably boisterous from the gin that flowed freely in that notorious era, often jammed the narrow lane, while the Cock Lane Ghost became the satirist's byword for credulity.

Confronted with these serious allegations of misdemeanour, as well as with public hysteria and the sensational reporting of the press, the Lord Mayor of London ordered an inquiry. On the investigative committee was Samuel Johnson, and after they examined Betty on 1 February, Johnson concluded that it was 'the opinion of the whole assembly, that the child has some art of making or counterfeiting a particular noise, and that there is no agency of any higher cause'. Once further tests showed that Scratching Fanny failed to make her presence known when Betty was physically restrained, the evidence was irrefutable. Richard Parsons, as the presumed instigator of the whole affair, was put on trial, found guilty of conspiracy and fraud and sentenced to be pilloried and imprisoned for two years.

The scandal of the Cock Lane Ghost fuelled ridicule of public superstitions like this: the case became a shorthand for trickery and its exposure. David Garrick, a friend of Johnson's, seized on the fad for his 1762 play *The Farmer's Return*, where he mocked city folk for being even more credulous than country bumpkins. Johnson himself didn't escape mockery. Even though he had been sceptical of the Cock Lane haunting, he retained a general belief that spirits could return from the dead and he was lampooned cruelly in the poem *The Ghost* (1762) by his rival, the satirist Charles Churchill. The Cock Lane affair soon became part of London lore; Charles Dickens, whose interest in ghosts needs no introduction, mentions it in *Nicholas Nickleby*, *A Tale of Two Cities* and *Dombey and Son*.

Holy ghosts

The continued urge to believe in the spirit world may have been a paradoxical result of encroaching materialism: ghosts – and demons too – were a bastion against atheism. These invisible phantoms could be enlisted in the cause of natural theology. In the seventeenth century Robert Boyle advocated the investigation of demonic manifestations because if they were real, then so was God. That was the explicit subtext of the Scottish mathematician George Sinclair's *Satan's Invisible World Discovered* (1685), an account of sightings that would prove 'that there are Devils, Spirits, Witches, and Apparitions'. When Thomas Hobbes argued that ghosts were mere fancies and delusion, he was suspected of implying that the same was true of the Lord.

Yet when Boyle's contemporary, the Cambridge Platonist Ralph Cudworth explained that belief in ghosts encouraged belief in 'one supreme ghost', he was voicing a complicated recommendation, for Christian theologians had long wrestled with the idea of the Holy Spirit. What is this numinous, neglected member of the Trinity? Is it, as some early Christians argued, to be merely equated with God, or should one follow St Augustine by insisting that it is neither Father nor Son? The debate hasn't been settled even now; according to the Scottish theologian John McIntyre, current views on the Holy Spirit are 'almost unrestricted'. A non-Christian might be forgiven for concluding that this is because the notion is too vague to be intelligible. But perhaps such vagueness is a strength, giving this invisible presence a Protean character that can be adapted to every age.

At face value, the Bible is of little assistance: it doesn't offer a unique doctrine of the Spirit, but on the contrary supplies grounds for several differing interpretations. The words used in both the Hebrew (*ruach*) and Greek (*pneuma*) translations can mean many things, all of them intangible and invisible: breath, air, wind or soul. When God breathes life into humankind in Genesis, the Holy Spirit might be considered to enter into man as a divine gift of life. This all-pervasive life-giving agency is how Hegel depicted the Spirit, using the word *Geist* that makes a link with ghostliness explicit. But the twentieth-century theologian Karl Barth worried that this association weakened any modern feeling for the Spirit:

The suspicion is solidified that the Holy Spirit is a phantom, a ghost. But common sense knows, especially since the Enlightenment, that there are no such things as ghosts.

As we will see, he was too optimistic about that.

The relative neglect of the Holy Spirit by theologians, compared to the other, more easily personified members of the Trinity, might stem primarily from this lack of a clear scriptural meaning or function. But it's possible too that we have simply lost the almost pantheistic sensibilities of the Bible's authors. The Holy Ghost is certainly more alive in modern Pentecostal churches, which have a stronger notion of being filled by ('baptized in', one might even say possessed by) the Spirit, than it is in today's staid tea-and-scones Anglicanism.

Some modern efforts to find a meaningful interpretation of the Spirit have sought to mine scientific thought, drawing analogies with the invisible force fields of physics. Those connections have sometimes been asserted rather too literally. The German theologian Wolfhart Pannenberg, who was influenced by Barth and Hegel, had the implausible hope that what one might call a 'field theory of the Holy Spirit' would unite science and religion, while John McIntyre offers a dubiously mechanistic picture which sounds more akin to Victorian pseudo-scientific explanations for telepathy. Just as sound waves stimulate the auditory system in a manner perceived as meaningful words, so 'it is conceivable that He [the Spirit, which McIntyre humanizes emphatically] may stimulate the brain in such a way as to enlighten the mind and so achieve the end-product of communication'.

A more fertile view recasts the Spirit as the *spiritus mundi* of the Neoplatonists, tailored to the era of ecological holism. The German theologian Jürgen Moltmann calls it 'the life force immanent in all the living, in body, sexuality, ecology, and politics.' This vision is most explicit in the present-day Green Christianity of American theologian Mark Wallace, who asks:

Could it be that the most compelling response to the threat of ecocide lies in a recovery of the Holy Spirit as a natural, living being who indwells and sustains all life-forms? Could it be that an earth-centred re-envisioning of the Spirit as the green face of God in the world is the best grounds for hope and renewal at a point in human history when our rapacious appetites seemed destined to destroy the earth?

We can see, then, that Barth was wrong. It is not the idea of a Holy Ghost that has suffered in recent times, but that of God the Father – too embodied an entity to appeal to any but the most literalist of believers. God has now Himself become the Spirit: disembodied, omnipresent, a life force and a *process* congruent with a contemporary view of the 'sacredness of the earth'. Those Baroque images of a radiant greybeard among the clouds now seem quaint if not absurd. God is not dead, he has just become invisible.

Fairy tales

Renaissance Neoplatonism bequeathed a secular view of ghosts and spirits: a belief that there might be a natural order of invisible beings that are neither souls nor demons and angels. Folk tradition had long maintained the existence of elusive spirit-like creatures and some natural magicians incorporated these into their world-view: Paracelsus and Agrippa described magical menageries that were expressions of nature's innate fecundity. This created a new category of spectres: as the English writer Joseph Glanvill, an apologist for the Royal Society, argued, they might be merely 'intelligent Creature[s] of the invisible World'. There were many names for these beings; some called them fairies.

In the Middle Ages fairies and sprites were neither small nor necessarily kindly – they were no fey, bloodless little Tinkerbells suitable only for children. Oberon was a demon summoned by necromancers, Robin Goodfellow (the forerunner of Puck) a household goblin in need of constant propitiation. Ariel in *The Tempest* is an elemental barely tamed by Prospero's magic, able to become invisible at will, and fundamentally a part of the enchanted isle's populous unseen world. It was only in the seventeenth century that such creatures became the 'little people', dwelling in woods and earth barrows and ancient places.

Paracelsus offered a detailed account of elemental beings in his book *On Nymphs, Sylphs, Pygmies and Salamanders*, probably written in the 1530s. In character they are just like us, he said, 'witty, rich, clever, poor, dumb':

[They] eat and enjoy the product of their labour, spin and weave their own clothing. They know how to make use of things, have wisdom to govern, justice to preserve and protect.

They differ from us in one key respect, however, for they have no souls – unless, that is, one of them persuades (or deceives) a human into marriage. That's why, Paracelsus claimed, these beings 'woo man [and] seek him assiduously and in secret'. Marriage between human men and female nymphs, he said, was not uncommon – although it was not always a happy union.

There are four types of these elementals, each of which inhabits one of the classical elements. Nymphs (or undines) dwell in water, sylphs (or sylvestres) in air, pygmies (gnomes) in the earth, and salamanders in fire. They guard the treasures of nature, ensuring that man does not rapaciously plunder them all. Are these beings truly invisible? Better to call them liminal: they exist between worlds. Paracelsus doesn't claim to have seen one himself, yet he knows what they look like. Gnomes in particular were occasionally glimpsed by miners: even the pragmatic expert on mining Georg Agricola says in *On Subterranean Animals* (1549) that they sometimes assist and encourage miners but at other times will throw pebbles at them (for which miner had not experienced the discomfort of a stone or two dropping on his head?).

This view of invisible elementals still had currency in the late nineteenth century. An Italian psychic named A. Farnese described them in *A Wanderer in the Spirit Lands* (1896), allegedly the autobiography of a man named Franchezzo whose spirit dictated it to the book's author:

Some are in appearance like the gnomes and elves who are said to inhabit mountain caverns. Such, too, are the fairies whom men have seen in lonely and secluded places. Some of these beings are of a very low order of life, almost like the higher order of plants, save that they possess independent motion. Others are very lively and full of grotesque, unmeaning tricks . . . As nations advance and grow more spiritual these lower forms of life die out from the astral plane of that earth's sphere, and succeeding generations begin at first to doubt and then to deny that they ever had any existence.

Mischievous imps and fairies that interfere in domestic matters were a stock of folk tradition, and if these beings were not necessarily invisibly small, their diminutive stature enabled them to avoid being seen nonetheless. You needed to watch out for them: they would nip, pinch and torment housewives and maids, knock over buckets and steal milk, although if placated with offerings of food and linen they would help with domestic tasks and leave money in shoes. In this way they made sure servants did their chores diligently. Ben Jonson attributes these vengeful habits to the Queen of the Fairies in his 1603 divertissement *The Satyr*:

> She, that pinches country wenches
> If they rub not clean the benches,
> And with sharper nails remembers,
> When they rake not up their embers.

Unseen fairies and goblins enforced other obligations too. Country lasses who were unchaste had better watch out for their taunts and pinches. They discouraged neglect of children, who might be replaced by changelings if not properly attended. By the same token they offered a useful excuse: I spilt the milk because a fairy tripped me; my child is defiant, ugly or retarded because he is a changeling.

You upset them at your peril. An account in an Irish newspaper in 1866 tells how, when a 'fairy stone' was inadvertently used in building a house, the dwellers were constantly bombarded with stones by invisible assailants. A West Sussex woman who used 'pixie stones' to make a rock garden was haunted by a tiny grey lady sitting on her rockery in silent reprimand, until the locals prevailed upon the woman to replace the stones before misfortune was visited on the whole village.

Despite occasional manifestations of this sort, the framework of fairy lore depended on invisibility, which denied any opportunity for proof or disproof. The common man, says Keith Thomas, 'knew that he could never count on actually *seeing* the fairies himself, for the little people were notoriously jealous of their privacy and would never appear to those who were so curious as to go looking for them'. In this way, says Thomas, 'there was an impenetrability about fairy-beliefs which protected them from easy exposure'. It also made them a gift to tricksters. The late-seventeenth-century aristocrat and Whig politician

Goodwin Wharton was strung along for years by his mistress Mary Parish, who sustained her comfortable affair with this dignitary by keeping him in touch with the fairy realm. Wharton never actually set eyes on the little folk (outside of his erotic dreams), but Parish assured him that fairies had a way of beckoning to those they wished to attract that was 'so quick . . . that none but those for whom it was intended could see it'. Fairies are hard to see anyway because they are so small, Parish explained – or rather, they are human size but wear special breastplates that act as a 'shrinking lens', an explanation that, if nothing else, indicates Parish's inventiveness.*

By the twentieth century these sprites were becoming children's fare, having been relentlessly sentimentalized by middle-class Victorians. When J. M. Barrie wrote in *The Little White Bird* (1902), the precursor to *Peter Pan*, that 'It is frightfully difficult to know much about the fairies, and almost the only thing known for certain is that there are fairies wherever there are children', he was putting them in their place rather than reflecting their tradition. The prettified 'flower fairies' of Mary Cicely Barker two decades later banished the last vestiges of their spite and caprice: now fairies *were* children, with Pre-Raphaelite locks and butterfly wings, flitting among the daisies.

But even then they had some adult advocates. In 1922 Arthur Conan Doyle told of 'Mr. Tom Charman, who builds for himself a shelter in the New Forest and hunts for fairies as an entomologist would for butterflies'. There are still shepherds on the South Downs, he added, who throw a bit of their bread and cheese over their shoulders at dinnertime 'for the little folks to consume'. All over the United Kingdom, he wrote, 'and especially in Wales and Ireland, the belief is largely held among those folks who are nearest to Nature'. And, he might have added, among romantics such as W. B. Yeats, whose belief in leprechauns seems to have been genuine. And not to forget mystics such as the Theosophist Charles Leadbeater, who expounded at length on fairies in *The Hidden Side of Things* (1913), where he claimed that they are of two types. There are 'elementals' that are mere 'thought-creatures' of the angels who govern the plant kingdom, which come

---

* Mary Parish evidently knew plenty about magical invisibility. Wharton's meandering autobiography records that she knew how to grow peas to give them an astrological virtue conferring invisibility when placed in the mouth – an obvious adaptation of traditional spells involving beans (page 9).

temporarily into being whenever 'one of these Great Ones has a new idea connected with one of the kinds of plants or flowers which are under his charge', and which dissolve when their work is done. And there are 'real nature spirits' which may be seen buzzing round flowers like humming birds when they have been 'ensouled' with 'etheric bodies'. Leadbeater uses the same terms as Paracelsus, saying that these beings will evolve eventually into salamanders (fire spirits) and sylphs (air spirits). Those of different nations have (rather fetchingly, like their football teams) different appearances: emerald-green in England, scarlet and gold in Sicily, deep blue shot with silver in New Zealand.

These scarcely visible little folk were further domesticated in children's tales such as Mary Norton's *The Borrowers* (1952). Yet Norton was alert enough to tradition to give her little household brownies the function of 'rationalizing' domestic mishaps and disappearances, like the vanishing of thimbles and pins – and to preserve the possibility that they might not exist at all, at least in the sober minds of adults. The equanimity of the Borrowers is wholly contingent on their *not being seen*:

> Pod looked at her blankly. 'I been "seen"' he said . . .
> Pod stared at Homily and Homily stared at the table. After a
> while she raised her white face. 'Badly?' she asked.
> Pod moved restlessly. ' I don't know about badly. I been "seen".
> Ain't that bad enough?'

Capturing the invisible folk

This 'protection from easy exposure' was (if the pun may be excused) as much promoted as it was challenged by the claim that fairies could be captured on photographic film. After all, by the twentieth century photography was well known to reveal phenomena invisible in the ordinary world, such as X-rays and radioactivity. So why not unseen beings?

What seems most disappointing about Arthur Conan Doyle's famous disgrace over the 'Cottingley Fairy' photographs is not that the creator of the arch-rationalist Sherlock Holmes was so easily hoodwinked but that he fell for such a horribly twee vision of what fairies are. Even one of the Spiritualists whom Conan Doyle consulted had the perspicacity to express scepticism at these 'Parisian-coiffed'

The Cottingley fairies, photographed in 1917.

sprites. Conan Doyle defended himself against that complaint in *The Coming of the Fairies* (1922): 'If they are conventional, it may be that fairies have really been seen in every generation, and so some correct description of them has been retained.' He didn't consider the more likely, and indeed the correct, explanation for their familiarity, as confessed by the perpetrator Elsie Wright towards the end of her life: that they were copied onto cardboard cutouts from a children's book.

Wright lived with her family in the picturesque village of Cottingley, near Bingley in Yorkshire, where as a teenager she produced the photographs, showing her and her younger cousin Frances Griffiths interacting with the elves, between 1917 and 1920. Conan Doyle, whose interest in the paranormal was well known to his acquaintances, was informed of the pictures by his friend Edward Gardner, a leading Theosophist, and Conan Doyle revealed them to the public in an article in the *Strand Magazine* in 1920.

'Short of final and absolute proof', he wrote, 'I consider, after carefully going into every possible source of error, that a strong *prima-facie* case has been built up.' He admitted that 'The cry of "fake" is sure to be raised' – but only, he insisted, by those 'who have not had the opportunity of knowing the people concerned'. He dismissed the 'stale and rotten' argument that stage magicians could manufacture illusions like this by sleight of hand:

There are few realities which cannot be imitated, and the ancient argument that because conjurers on their own prepared plates or stages can produce certain results, therefore similar results obtained by untrained people [indeed, children] under natural conditions are also false, is surely discounted by the intelligent public.

He didn't seem to consider it pertinent that Elsie's father was a keen amateur photographer and that Elsie had begun to dabble in the hobby. Like so many dupes of that era, Conan Doyle felt himself both a sound judge of character and bound as a gentleman to regard gentility as a guarantee of honesty.

His argument for why the existence of invisible beings was consistent with modern science was by that time well rehearsed, as I shall consider in more depth in the next chapter:

We see objects within the limits which make up our colour spectrum, with infinite vibrations, unused by us, on either side of them. If we could conceive a race of beings which were constructed in material which threw out shorter or longer vibrations, they would be invisible unless we could tune ourselves up or tone them down. It is exactly that power of tuning up and adapting itself to other vibrations which constitutes a clairvoyant, and there is nothing scientifically impossible, so far as I can see, in some people seeing that which is invisible to others.

'It seems to me', he concluded,

that with fuller knowledge and with fresh means of vision these people are destined to become just as solid and real as the Eskimos . . . These little folk who appear to be our neighbours, with only some small difference of vibration to separate us, will become familiar. The thought of them, even when unseen, will add a charm to every brook and valley and give romantic interest to every country walk.

That 'charm' was of a decidedly cloying nature, if the account of a clairvoyant who accompanied Elsie to Cottingley Glen in 1921 is anything to go by. She might as well have been flicking through Barker's flower-fairy books and one senses Walt Disney sharpening his pencil:

In the field we saw figures about the size of the gnome. They were making weird faces and grotesque contortions at the group. One in particular took great delight in knocking his knees together . . . *A*

Brownie. He is rather taller than the normal, say eight inches, dressed entirely in brown with facings of a darker shade, bag-shaped cap, almost conical, knee breeches, stockings, thin ankles, and large pointed feet . . . A fairy with wings and general colouring of sea-blue and pale pink. The wings are webbed and marked in varying colours like those of a butterfly . . . There is a tinkling music accompanying all this.

But there was a deeper agenda driving Conan Doyle's gullibility. For all Sherlock Holmes' dispassionate logic, his creator felt a keen sense of spiritual loss in the disenchantment of the modern world:

> The recognition of [the fairies'] existence will jolt the material twentieth-century mind out of its heavy ruts in the mud, and will make it admit that there is a glamour and a mystery to life. Having discovered this, the world will not find it so difficult to accept that spiritual message supported by physical facts which has already been so convincingly put before it. All this I see, but there may be much more. When Columbus knelt in prayer upon the edge of America, what prophetic eye saw all that a new continent might do to affect the destinies of the world? We also seem to be on the edge of a new continent, separated not by oceans but by subtle and surmountable psychic conditions. I look at the prospect with awe.

Although Elsie and Frances never recanted on their claim to have *seen* real fairies, they did admit in the early 1980s that the photographs themselves were faked. They said that the affair began as 'a bit of fun', but that they quickly became too embarrassed to admit to it. 'Two village kids and a brilliant man like Conan Doyle – well, we could only keep quiet,' said Elsie. Frances confessed to being puzzled at how such serious thinkers were so easily deceived, although in fact she surely intuited the truth of the matter: 'they wanted to be taken in.'

All the same, for their time and with the resources available the Cottingley fairy photos were faked with extraordinary skill. But the public appetite for them is all the more surprising given that the same cycle of fascination and awe followed by discredit and disillusionment had only just run its course with ghost photographs.

The editor of *Waverley Magazine* Moses A. Dow, photographed by William Mumler around 1871 with the spirit of Dow's assistant and adopted daughter Mabel Warren, who died in 1870.

Acceptance of that piece of trick photography is easier to understand, for the spirit pictures became popular before it was widely known how easily they could be concocted. Their origins might have been genuine enough: chemical reactions on the surface of glass plates used for the early 'wet collodion' photographic technique could preserve faint images of an earlier exposure that would show up, probably to the alarm of the photographer, when the plates were reused.

An American engraver and photographer named William Mumler is generally credited with the first ghost photograph, an image of his dead cousin that he published in 1862. Mumler set up a 'spirit photography' business in Boston and New York, but narrowly escaped charges of fraud in 1869 when it was found that one of the 'spirits' in his images was still alive. The showman and paranormal debunker P. T. Barnum testified against Mumler in the ensuing trial, where he challenged

Mumler's celebrated image of Abraham Lincoln's wife with the 'ghost' of her dead husband.* To illustrate Mumler's fraud, Barnum commissioned a photographer to fake his own portrait alongside Lincoln. Mumler escaped conviction, but his business was ruined. David Brewster had already explained in the 1850s how these ghostly images could be easily manufactured because of the long exposure times then needed to take photographs.

Such explanations did little at first to diminish the popularity of the genre, for in its mysterious ability to capture the instant and to solidify intangible light photography seemed virtually a supernatural medium already. (The French word *séance* could also refer to a photographic sitting.) Didn't it, after all, convey a weird kind of immortality – and paradoxically, by doing so, remind the sitter that death awaited? Because photographs survive us, says cultural theorist Eduardo Cadava, 'the photograph is a farewell'. It hardly seemed to stretch credibility, therefore, to suggest that the photographic process could reach beyond that departure of the soul. In 1872 Frederick Hudson took a photo of the medium Elizabeth Guppy and her husband and 'found' a veiled spirit also in the frame. He teamed up with a 'painting medium' named Georgiana Houghton to sell reproductions of their ghost photos, and in 1882 Houghton published *Chronicles of the Photographs of Spiritual Beings and Phenomena Invisible to the Material Eye*, a collection of many such photographs of spirits and ghosts. In his 1898 manual of stage magic, Albert Hopkins described again how such images could be fabricated, commenting archly that Houghton's ghost photographs often suffered 'accidents' when the elusive spirits failed to reveal themselves.

Photography was devised in 1839 by Louis Daguerre in France and William Henry Fox Talbot in England, and from its earliest days it seemed to be as much about revealing the invisible as documenting the visible. Nathaniel Hawthorne's novel *House of the Seven Gables* (1851) included a maker of daguerrotypes within which people's true nature was revealed, however they might appear in the 'real' world.†

* The Lincolns were enthusiasts of Spiritualism, and were said to have conducted séances in the White House.
† The camera, it seems, never lies about spirits. This putative ability of photographs to disclose what is hidden has survived into the digital era, as for example in the

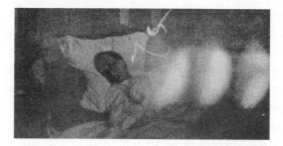

Hippolyte Baraduc took this photograph of his own wife 20 minutes after she died in 1907. It allegedly shows her soul departing as a white misty cloud.

In the 1860s Carl von Reichenbach attempted to prove his hypothesis of odic rays by recording their imprint on photographic plates. The advent of X-ray photography in 1896 (see page 92) seemed to confirm this role. In his book on 'psychic photography' *Photographing the Invisible* (1911), James Coates considered it a scientifically proven fact that photography can fix in its emulsion with equal facility 'the visible, the material invisible, and the immaterial invisible or the psychic.' 'To say that the invisible cannot be photographed', Coates asserted, 'would be to confess ignorance of facts which are commonplace.' The French neurologist Hippolyte Baraduc claimed to have found a form of photography that could even reveal the human soul, while he and others imagined they could capture thoughts and dreams this way. Another Frenchman, Commandant Louis Darget, developed a Portable Radiographer, a device for strapping a photographic plate to the forehead where it registered thoughts (conveyed by invisible rays Darget called V-rays) as vague streaks and blobs. Darget made fanciful interpretations of these indeterminate shapes: a kind of whitish rhombus 'received' from a pianist while playing the piano and contemplating a bust of Beethoven looked (Darget implausibly insisted) like the composer himself.

Thought photography was endorsed enthusiastically by Annie Besant and Charles Leadbeater in *Thought Forms* (1901), although they said that only a truly sensitive human medium, rather than the chemical medium of a photographic plate, could truly resolve these forms. The English novelist Sax Rohmer, who claimed to be a Rosicrucian and a

disclosure of a demonic possession in the final moments of the 2011 movie *Insidious*: on the digital screen of the exorcist's camera, the lead character is suddenly revealed as a grotesque gargoyle.

member of the Hermetic Order of the Golden Dawn, invoked thought photography (transmitted by 'the odic force, the ether – say it how you please') as one of the techniques used by Moris Klaw, hero of *The Dream-Detective* (1920), to solve mysteries by occult means. One wonders whether the colourful but mysterious 'thought patterns' now revealed by magnetic resonance imaging of brains are liable to meet the same fate.

Tricks of the light

It is quite natural that one of the first uses of photography would be to make invisible beings visible. For optical technology has always been closely allied with magic and was long thought capable of revealing what went otherwise unseen – particularly spirits, souls and demons, or something like them. In his *Occult Philosophy* Agrippa wrote that

> [b]y the artificialness of some certain Looking-glasses, may be produced at a distance in the Aire, beside the Looking-glasses, what images we please, which when ignorant men see, they think they see the appearances of spirits, or souls; when indeed they are nothing else but semblances kin to themselves, and without life.

The camera obscura, in a form that artists could use to copy the projected image. From Athanasius Kircher's *Great Art of Light and Shadow* (1646).

Agrippa might here be referring to the camera obscura, the forerunner of the photographic camera, in which natural scenes are projected through a small opening into a darkened space. Early observers of the camera obscura must have felt deeply unnerved by this conjuration. The image appears upside down, but in *Naturalis Magiae* (1558) Giambattista della Porta described how to right it using a mirror. By the early seventeenth century mountebanks were using such devices to astonish audiences. Investigations of optics were especially liable to attract accusations of witchcraft, because of the strange, disorientating effects they produced.

Looking-glasses that produce figures 'at a distance in the air' also featured in the magic lantern, an early form of projector that became a stalwart device of optical natural magic. It was described by the Jesuit inventor and mystical philosopher Athanasius Kircher in his *Great Art of Light and Shadow* (1646): light is passed through an image painted onto glass and then through a lens before falling onto a screen. By the time Kircher was writing, magic lanterns were becoming commercialized. Samuel Pepys, always on the lookout for new scientific toys with which to divert himself, bought one from the London optical instrument maker John Reeves in 1666, along with 'pictures

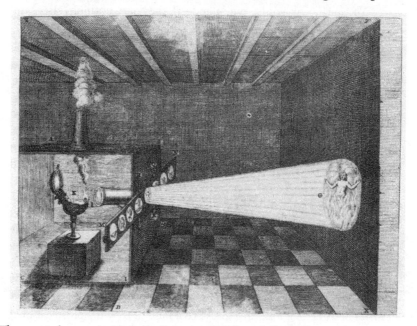

The magic lantern in Kircher's *Great Art of Light and Shadow*.

on glass, to make strange things appear on a wall'. He pronounced it 'very pretty', but others found it frightful. The Danish mathematician Thomas Walgensten travelled across Europe selling these lanterns and using them purportedly to summon ghosts.

Although the magic lantern was only perfected in the mid-seventeenth century, possibly by the Dutch scientist Christiaan Huygens, there are a few records of similar devices from much earlier. Right from the start they were used to conjure up demons and other grotesque apparitions. The early fifteenth-century Italian physician Giovanni Fontana implied that the magic lantern might be used as a military weapon to terrify the enemy. While he, and later Walgensten, may have been upfront about the illusionism involved, less scrupulous individuals used these instruments to demonstrate supposedly magical and necromantic powers. They became a part of the arsenal of tricks used by intinerant tinkers and rogues who plied the country fairs and lanes beguiling common folk. 'In the seventeenth century', says folklore historian Owen Davies, 'the ability

A magic lantern projecting a demon, from Giovanni Fontana, *Bellicorum instrumentorum liber* (c.1420).

PHANTASMAGORIA,
THIS and every EVENING,
AT THE
LYCEUM, STRAND.

This advertisement for Paul Philidor's Phantasmagoria show in London in 1801 shows that ghosts were a central element of the performance.

of magic lantern operators to conjure up such spirit images led to a blurring of the boundaries between theatre, necromancy and natural magic.' Light became a medium for both revealing and creating invisible beings.

The magical stage spectacles of the late eighteenth century straddled this ambiguous boundary. The German illusionist Johann Georg Schröpfer held séances in his Leipzig coffee shop in which he used the magic lantern, projected onto smoke, to summon ghosts. Schröpfer's performances were perhaps the first 'entertainment séances', but he was secretive about the mechanics of his conjurations and, as a Freemason with a genuine interest in the occult, he seemed somewhat uncertain of their status himself. Legend has it that he shot himself after becoming convinced of the reality of his own illusions. His techniques were copied by the German Paul Philidor, whose popular public displays in the early 1790s were unashamedly eye-catching and became known as 'phantasmagoria'.

Subsequently, Étienne Gaspard Robertson used magic-lantern back-projection in his 'Fantascope' shows, in which, by mounting the device on wheels, he could make the projection grow rapidly larger or smaller so that ghouls and demons might seem to rush upon the terrified audience. Robertson explicitly sought to scare his public with these visions: he was in effect producing the first horror films. He introduced an illusion of movement in the projected images by superimposing

An eighteenth-century engraving of one of Robertson's magic-lantern light shows.

slides, for example with disembodied eyes that made a face appear to glance this way and that. Robertson was a professor of physics with a special interest in optics, who realized the commercial potential of optical trickery when he attended one of Philidor's extravaganzas. And although he made no pretence of possessing magical abilities, he exploited his specialist knowledge while artfully keeping his audiences guessing about what they were seeing. According to Barbara Maria Stafford, his shows created an atmosphere of 'scientific necromancy', an 'unholy marriage between an ancient thaumaturgy and a modern, sensational physics.'

The most famous illusionistic ghost of the stage also comes from this collusion of science demonstration and pure theatre. In the mid-nineteenth century, the Royal Polytechnic Institute in London put on magic and séance shows to show how paranormal activities could be faked. One of the lecturers was the chemist and science popularizer John Henry Pepper, who later set up his own 'Theatre of Popular Science and Entertainment' at the Egyptian Hall and took his show on tours of the United States and Australia. Pepper collaborated with the Liverpool engineer Henry Dircks in the late 1850s to create a technique for projecting the reflection of a hidden actor onto a huge, slanted sheet of glass: a semi-transparent apparition perfect for depicting ghosts. Plays featuring 'Pepper's ghost', including *Hamlet*, *Macbeth* and *A Christmas Carol*, became sensations throughout Europe

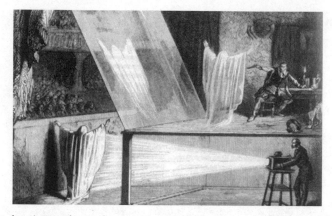

'Pepper's ghost' was a device that summoned ethereal spirits to the nineteenth-century stage.

and the United States, although the costly and cumbersome set-up, with a glass wall placed on the stage, limited the use of the illusion.

Pepper's ghost triggered something of a revival (if that is not a singularly inapt expression) for Shakespeare's ghosts, for their literal depiction by an actor was considered crude by the middle of the eighteenth century. They had vanished from sight, to be summoned only by the exaggerated responses of the actors, like David Garrick, who 'saw' them. Embodied ghosts became fashionable again in the age of Gothic romanticism, but theatre audiences could be slow to accept them: they hissed during a performance of *Macbeth* in Covent Garden in 1807 when Banquo was played by an actor. Pepper's ghost was, save for its mechanical complication, perhaps the ideal compromise: an apparition that moved but remained insubstantial, on the border of the unseen. The debate over whether or not Shakespeare's ghosts should be visible continues today: some recent Hamlets have played to an unseen and thus 'psychological' ghost, others to actors or to wraiths rendered with light, shadow and trick photography.*

The elaborate illusionism of the theatrical light-show found a new home in the early days of cinematography. In the late 1880s Thomas Edison began work on a kind of electrical magic lantern called the Kinetoscope that projected a series of still images in rapid succession

---

* In *The Haunted Hotel* (1878) Wilkie Collins has a playwright toy with the idea of evoking a ghostly presence using smell alone, but he fears that '[he] shall drive the audience out of the theatre'.

to create the illusion of movement. In 1894 he opened a Kinetoscope parlour in New York, where for a few cents one could watch the first motion pictures, each lasting a minute or so. But these viewing devices were relatively immobile, purpose-built units on which only one person at a time could watch the movie through an eyepiece. Determined to outdo their American rival, the Lumière brothers Auguste and Louis in France turned the magic lantern into a portable, manually operated movie projector called the *Cinématographe* that threw the image onto a screen. A paying Parisian audience watched the first public screening in 1895. The subject of that brief film – workers leaving the Lumière factory – seems deeply underwhelming today. But given the genealogy of these devices, it is no surprise that marvels soon took over. Illusions, ghosts in particular, were popular subjects for the pioneers of the motion picture.

The Frenchman Georges Méliès was captivated by stage magic when, working in London as a clerk in the mid-1880s, he went to the Egyptian Hall to see the illusion shows of John Nevil Maskelyne, who succeeded Pepper as the resident magician. Back in Paris, where his father forced him to work in the family's shoe business, Méliès took to frequenting the Théâtre Robert-Houdin to watch Jean Eugène Robert-Houdin himself. In 1888 he used the money from his share of his retired father's business to buy the theatre. Here Méliès developed many ingenious new illusions, including magic-lantern displays, and he incorporated them into fantastical narratives to create a truly magical form of theatre.

Méliès attended the premiere of the Lumière brothers' *Cinématographe*, after which he bought a movie camera and started making films himself. Many of these used his existing stage tricks, supplemented by the new illusionistic possibilities that photography offered. He made 78 films in 1896 alone, and over 500 during the next two decades: strange, beautiful visions, to which Martin Scorsese's *Hugo* (2011) pays reverent homage. Several of them were ghost films; some had a macabre edge, but others treated the subject matter humorously. The invisibility of spirits always had comic possibilities, but Elizabethan and Jacobean farce rarely exploited them: ghosts were generally too serious a matter for that. Méliès' *The Apparition, or Mr Jones' Comical Experiences with a Ghost*, in contrast, shows the hapless Jones plagued by an unseen spirit who moves around his hotel

A scene from Georges Méliès' *Le Manoir du diable* (*The Haunted Castle*) (1896).

furniture like a poltergeist version of Buster Keaton. Seeing ghosts do funny things at the movies couldn't but undermine their capacity for instilling fear. The trope has proved enduring, whether it is a white-coated, clownish Hopkirk, invisible to his foes as he aids his detective partner Randall from beyond the grave in the 1970s TV series *Randall and Hopkirk (Deceased)* or the antics of Bill Murray and Dan Ackroyd in *Ghostbusters* (1984).

The English inventor and psychic George Albert Smith, a member of the Society for Psychical Research (see page 136), made several ghost films in the 1890s which used special effects such as 'stop motion' to make objects and people vanish and materialize, or double exposure to summon partly transparent ghosts. Edison was quick to discern the genre's appeal, releasing *Uncle Josh's Nightmare* in 1900, in which objects and phantoms appear and vanish to the discomfort of poor Uncle Josh. Literary ghosts were quickly adapted for the screen: there were four movie versions of *A Christmas Carol* between 1901 and 1913 and a *Macbeth* in 1916. Nor was a fascination with invisibility among early cinematographers confined to ghosts. Cecil Hepworth, whose pioneering short film of *Alice in Wonderland* (1903) played with bizarre illusionism, made the comedy *Invisibility* (1909), while Pathé's *L'Homme invisible* (1909; the identity of the director is disputed) was explicitly inspired by H. G. Wells' eponymous novel. Aleister Crowley, practising a magic ritual of invisibility during his travels in Mexico, seemed to intuit the link between strange optical effects and the cinema:

I reached a point where my physical reflection in a mirror became faint and flickering. It gave very much the effect of the interrupted images of the cinematograph in its early days.*

In such ways, films of ghostly and supernatural phenomena were not simply an early genre of cinema – they were its natural subject, for the motion picture should properly be seen not so much as 'celluloid theatre' but as celluloid magic. For its first audiences, cinema was simply an extension of stage magic, with all its associations of marvellous vanishings and eerie manifestations. Didn't even the abrupt appearance of images in a darkened room, accompanied by the purr of the projector, have the atmosphere of a séance? The magician and escape artist Harry Houdini made films of his exploits; as film scholar Matthew Solomon comments, '[t]he cinematic medium perfected the dematerialization of the body that had been one of the magician's specialities'. What is more, it immortalized the actors on the screen. After seeing the Lumière brothers' first films in 1896, the French journalist Jean Badreux wrote that they

> have found a way to revive the dead . . . in a word, we will be able to bring those who are no longer in this world back to life before our very eyes. Science has triumphed over death.

That same year, the Russian writer Maxim Gorky saw the Lumière films in Nizhny-Novgorod and wrote that '[i]t seems as if these people have died and their shadows have been condemned to play cards in silence into eternity.' Jacques Derrida seemed to discern this character of cinema when he called it 'the art of ghosts, a battle of phantoms'. (That, at least, he opined with straight-faced impishness, is 'what I think the cinema's about when it's not boring'.)

The tradition of magical illusion remains as strong as ever in the dominance of special effects in popular cinema: movie-makers are, as Steven Spielberg's company reminds us, purveyors of industrial light

---

* Crowley asserted that he eventually mastered the trick to which Eliphas Lévi alluded of 'prevent[ing] people noticing you when they would normally do so.' As a result, he says, 'I was able to take a walk in the street in a golden crown and a scarlet robe without attracting attention.' No one could accuse Crowley, and several others of his Golden Dawn crowd, of lacking in grandiosity.

and magic. And as for spooks, an astute director knows that the chills are maximized by what is unseen. When the demands of Hollywood spectacle overwhelm the power of suggestion, the result is always disappointing: the hellish beast in the climax of Tobe Hooper's *Poltergeist* (1982) is worthy of Méliès, but not in a good way. The invisible presences in *The Blair Witch Project* (1999) and *Paranormal Activity* (2007) are much more terrifying – audiences of the latter allegedly fled the cinema in terror, and the movie's advertising took the bold step of showing viewers' hysterical reactions to screenings. Perhaps, then, Garrick had a point: keep your ghosts invisible, and they will be all the more terrible.

Modern ghosts

The fear induced by human souls when manifested as invisible beings is not, however, universal. Although ghosts have long been a source of terror in the West, the unseen presence of departed ancestors is taken for granted in China and Japan, where it seems at times almost like a cosy domestic arrangement. Their presence is palpable even if their form is invisible – a place is laid for them at the supper table and they are served first. China's Hungry Ghost Festival on the 15th night of the seventh lunar month is no chilling Hallowe'en but a celebration in which fake food, money and other items are offered to the ancestors. At live performances during the festival, the first row of seats is reserved for the invisible ghosts, and this opening of the gates of the underworld to release its unseen souls has the air of a bustling open day. Perhaps Keith Thomas unwittingly captures the difference between Eastern and Western ghosts in his remark that 'the main reason for the disappearance of ghosts [in the West] is that society is no longer responsive to the presumed wishes of past generations'.

But have ghosts really disappeared, or have they just taken on new forms? Belief in ghosts seems more susceptible to fashion than to attrition in the face of scientific rationalism. A third of the British population attests to such convictions today, compared to a tenth in the 1950s, which Owen Davies attributes to their increased social acceptability. Half the population of the United States believes in ghosts, but since more than two-thirds believe in the Devil this should come as no surprise.

The more complex matter is what these professed beliefs amount to. Some people might credit the existence of lonely Gothic ghosts who wander forlorn in country houses, eerie but harmless. Others prefer the malicious, murderous phantoms of modern horror films – a type perhaps first imagined in Fitz-James O'Brien's short story 'What Was It?' (1859), in which the narrator is assailed by an invisible but very palpable creature, more akin to an evil elemental. 'There it lay, pressed close up against me, solid as stone', he says, 'and yet utterly invisible!' His friends help him to restrain and bind the unseen thing and once they have sedated it with chloroform, the men take a plaster cast of the body, which reveals a form 'distorted, uncouth, and horrible, but still a man':

> Its face surpassed in hideousness anything I had ever seen. Gustave Doré, or Callot, or Tony Johannot, never conceived anything so horrible . . . It was the physiognomy of what I should have fancied a ghoul to be. It looked as if it was capable of feeding on human flesh.

This could almost be a screenplay for Hammer Films. Such nihilistic invisible demons signify nothing, existing only for their potential to thrill and terrify, to rend and slaughter like Freddy Krueger coalescing out of an unquiet dream.

Some of us still choose to regard ghosts as sombre visitations of the dead, like those conjured up by Victorian mediums. Many admit simply to having experienced things they cannot explain: weird sounds, objects moving, a sense of dread in a particular room. Modern ghosts might even be benign agents that console and right personal wrongs, as in the 1990 movies *Ghost* and *Truly, Madly, Deeply*. Ghosts today are demons or angels, all the more diminished by their explicit moral allegiances.

Arguably the first truly modern ghost was Sir Simon of Canterville Chase in Oscar Wilde's *The Canterville Ghost* (1887). No longer able to spook the materialistic American residents of the old country house, he is a tragicomic character, a spectre still all too human: sad, inept, frightened, accident-prone, even scared himself of a mock ghost rigged up by the children of the house. In fact he is merely a commodity, a part of the real estate he haunts – and as such, a potential tourist attraction. 'I reckon', says the head of the house, Mr Otis, 'that if there were such a thing as a ghost in Europe, we'd have it at home in a very short time in one of our public museums, or on the road as a show.' Once ghosts lose their religious and social motivation, they are apt to go the way of all non-human agencies in the public perception, whether they be animals, robots, aliens or gods: we make them just like us.

Yet ghosts always find new places to haunt. As we have seen already, the development of visual recording media multiplied them. In this respect the phonograph for recording sound, pioneered by Thomas Edison, was even more disturbing and fertile than the photograph – for people were long accustomed to portraits of the deceased, but sound and voice, so evocative of the person' physical presence, had previously been ephemeral. Here the human breath itself – *pneuma*, the spirit – seemed to become trapped in a machine. When Edison demonstrated his phonograph (a sheet of tinfoil wrapped around a cylinder onto which a needle inscribed grooves that recreated sound vibrations) in 1877, it seemed to present the possibility almost of bringing souls back from death. As one reporter wrote, 'There is many a mother mourning her dead boy or girl who would give the world could she hear their living voices again – a miracle your phonograph makes possible.' 'Death has lost some of its sting', claimed another in 1896, 'since we are able to forever retain the voices of the dead.' Remember Nipper, the little dog in the logo of His Master's Voice, forever obedient to – and forever mourning – his dead master, whose voice is immortalized on the phonograph record?

Voice recording seemed capable of summoning from their graves even those who had long perished. The American philosopher and logician Charles Sanders Peirce predicted that, given enough time, science 'may be expected to find that the sound waves of Aristotle's voice have somehow recorded themselves'. These sonic ghosts would be oracular, like the ghost in *Hamlet*: they would speak but not converse and we might struggle to comprehend these messages never meant for us. 'By preserving people's apparition in sight and sound', says media historian John Durham Peters, 'media of recording helped repopulate the spirit world. Every new medium is a machine for the production of ghosts.'

That was never in doubt in the case of radio, which spawned a fresh contingent of ethereal phantoms, of lost voices drifting from

HIS MASTER'S VOICE.

His Master's Voice: the dead master speaks to his dog.

the airwaves, dead lovers and soldiers slain in the Great War. This technology elicited dreams of tuning in to the voices of history. Marconi and Edison both professed to be investigating devices that would make electronic contact with the dead; Marconi hoped radio technology might somehow pick up the words of great figures of the past, perhaps even the last words of Christ on the cross.

How much more potent, then, when you could see electronic ghosts as well as hear them. In cinematography you could at least see the celluloid traces responsible for the apparition but in television the phantoms streamed through the ether. It might have seemed natural and harmless enough to refer to the double images of early television sets, caused by poor reception or synchronization of the electron beam, as 'ghosts' – but this terminology spoke to, and fed, a common suspicion that the figures you saw on the screen might not always correspond to real people. After all, they too might already be dead. News reporters flocked to the home of Jerome E. Travers of Long Island in December 1953 to witness the face of an unknown woman who had appeared on the screen and wouldn't vanish even when the set was unplugged. (The family had turned the screen towards the wall, as if it was in disgrace.) Some early television viewers suspected the figures on the screen could see them, and would refuse to undress in front of them. They suspected that more sinister forces than the smiling news readers lurked behind the monochrome glow. 'There is nothing wrong with your television set', insisted the 1960s series *The Outer Limits*. 'Do not attempt to adjust the picture. We are controlling transmission.' Who, though, was *we*? In the movie *Poltergeist* it was malevolent spirits, for whom the screen was a barrier permeable in both directions: the family's small daughter was sucked into the televisual ether, and all that remained was a disembodied voice wavering with the static of poor reception.

'I believe that ghosts are a part of the future', said Derrida in 1983, 'and that the modern technology of images like cinematography and telecommunication enhances the power of ghosts and their ability to haunt us.' So who could possibly be surprised that the internet throngs with ghosts – that, as Owen Davies says, 'cyberspace has become part of the geography of haunting'. Here too the voices and images of the dead may linger indefinitely; here too pseudonymous identities are said to speak from beyond the grave. More even than the telephone and television, the internet, that invisible babble of voices, seems almost designed to house spirits, which after all are no more ethereal than our own cyber-presence.

John Durham Peters is surely right, therefore, that all modern media are ghost factories, forever manufacturing what the nineteenth-century psychic researcher Frederic W. H. Myers called 'phantasms of the living': disembodied replicas of ourselves, ready to speak on our behalf. By appearing to transmit our presence over impossible reaches of time and space, and preserving our image and voice beyond death, these media subvert the laws that for centuries constrained human interaction by requiring the physical transport of a letter, or the person themself. We submit to the illusion that the breath of the beloved issues from the phone, that the Skyped image conjured on the screen by light-emitting diodes is the far-away relative in the flesh. We invent endless password-protected selves, from patchworks of photograph, video footage, fleeting remarks. Appearing 'in person' is now a subcategory of appearing at all. 'To interact with another person', says Peters, 'could now mean to read media traces.'

But this conversation with media phantasms, with recorded selves, is never enough. Kafka expressed it vividly: 'Written kisses don't reach their destination, rather they are drunk on the way by the ghosts.' And faster than we devise new ways to convey humans from here to there, we invent new ways to summon our media-incarnated doppel-gängers. In the process, 'the spirits won't starve', said Kafka, 'but we will perish.' As Peters writes,

> The concern in psychical research – contact with spectral emanations of distant bodies, whether via writing, images, sounds, or even touch – is part of a larger effort in modernity to reorganize representations of the human body.

Ever since modern media first developed as liminal agencies, these ghostly impressions have had a natural abode: an invisible realm that was once known, and colloquially is still, as the ether.

4 Rays that Bridge Worlds

Invisible things are the only realities.

William Godwin, *Mandeville* (1817)

'Is the invisible visible?'
'Not to the eye, but its results are.'

Wilhelm Röntgen,
interviewed for *McClure's Magazine*, 1896

When nineteenth-century science permitted a sliver of natural magic to remain lodged in its side, no one could have guessed where that would lead. This tenacious shard was the ether: a classical inheritance, being another name for Aristotle's fifth essence or quintessence that was thought to infuse the heavens. As such, ethereal fluids played a central role in natural magic. We saw earlier how a pervasive 'spirit of nature' that imbued all things with animation and vitality featured in the Neoplatonic cosmologies of Renaissance mages.

Emanations of an intangible nature were widely considered to convey the influence of the heavenly bodies to the earthly sphere. It was through these occult traditions that both 'ether' and 'spirit' became synonymous with any immaterial, invisible vapour escaping from a substance, and they survive in this guise in chemical parlance today. But for physicists the ether evolved into an invisible, subtle medium considered to be the bearer of light rays: it was luminiferous.

In the mid-nineteenth century James Clerk Maxwell showed that light is a wave of oscillating electric and magnetic fields – the two best attested occult forces of antiquity. The ether was judged to be the vehicle that carried them, all the way from the sun and the stars. Maxwell, a devout Presbyterian, spoke of this invisible, ubiquitous sea as something divine, in terms more or less indistinguishable from Isaac Newton's concept of the *Sensorium Dei*, the all-pervading presence of God throughout absolute space. Maxwell wrote that

> The vast interplanctary and interstellar regions will no longer be regarded as waste places in the universe, which the Creator has not seen fit to fill with the symbols of the manifold order of His kingdom. We shall find them to be already full of this wonderful medium; so full that no human power can remove it from the smallest portion of space, or produce the slightest flaw in its infinite continuity. It extends unbroken from star to star; and when a molecule of hydrogen vibrates in the dog-star, the medium receives the impulses of these vibrations.

Scarcely any scientist doubted that the ether existed. 'One thing we are sure of and that is the reality and substantiality of the luminiferous ether', attested William Thomson, later Lord Kelvin, in 1884.* Vestiges of natural magic were retained in this medium, even if Maxwell and others attempted to reduce it to something purely mechanical, a kind of springy fluid. Some scientists considered the ether to link the physical and the spiritual worlds, enabling invisible presences to communicate with us: it connected science to the supernatural. Perhaps (they thought) the ether might even be the very fabric of worlds both seen and unseen. Thomson speculated in 1867 that atoms might be knotted tubular vortices of ether – an idea

---

* Einstein's theory of special relativity, published in 1905, seemed to show that we could do without the ether. His description of light waves made the electromagnetic fields self-supporting, without the need for any other medium to carry them: there was nothing left for the luminiferous ether to do. 'The electromagnetic fields appear as ultimate, irreducible realities, and at first it seems superfluous to postulate a homogeneous, isotropic ether-medium and to envisage electromagnetic fields as states of this medium,' Einstein wrote in 1920. But, contrary to popular belief, even Einstein was not prepared to relinquish this useful concept, as the 'at first' here attests. 'There is a weighty argument to be adduced in favour of the ether hypothesis', he wrote. 'To deny the ether is ultimately to assume that empty space has no physical qualities whatever. The fundamental facts of mechanics do not harmonize with this view.' Einstein formulated a new ether theory based on his conception of space, time and gravity in the theory of general relativity.

Maxwell found appealing – while the Siberian chemist Dmitri Mendeleyev included 'ether', composed of etheric atoms, as one of the elements in his revised periodic table of 1903. In the late nineteenth century, matter itself seemed on the verge of being dematerialized.

New scientific discoveries made the idea of the ether as a bridge between worlds ever more plausible. In 1895 the German physicist Wilhelm Röntgen discovered an invisible form of radiation with tremendous penetrating power: unlike light, it could travel through black paper and other materials. This emanation could leave its trace imprinted in photographic emulsion, turning it dark where the rays struck and thereby making visible not only the invisible rays themselves but also the hidden structures they encountered en route. It can be no coincidence that in the age of Spiritualism the first hidden objects brought to light by these 'Röntgen rays' were the bones in the hand of Röntgen's wife: a presentiment of death captured on film, as she herself exclaimed fearfully. Even Röntgen's name for these rays, chosen at first as a provisional place-holder, seemed to carry connotations of something occult and ominous: X-rays.

X-rays joined a growing list of invisible rays around the turn of the century: cathode rays and anode rays (or canal rays), discharge

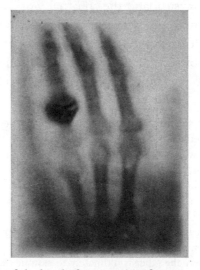

The X-ray photograph of the hand of Röntgen's wife, Anna, taken in his laboratory in Wurzburg at the end of the nineteenth century. 'I have seen my death!', she is said to have cried.

rays, uranic rays (radioactivity), black light, N-rays, cosmic rays and others. Many of the researchers involved in the early study of X-rays, cathode rays and radioactivity, including J. J. Thomson, Pierre Curie and the English scientists Oliver Lodge and William Crookes, had a profound, if sometimes sceptical, interest in séances and the paranormal. Science and Spiritualism fed one another. Science seemed on the verge of disclosing unseen realms of existence, while belief in spirit communication – a response to disillusionment with materialism and orthodox religion – created a receptive atmosphere for reports of unproven, invisible rays. 'We know little about the medium that surrounds us', the Curies admitted, 'since our knowledge is limited to phenomena which can affect our senses, directly or indirectly.'

For many scientists, the discovery of invisible rays bolstered a desire to recast old and deeply held beliefs within a scientific framework: they offered sustenance to spirits. They were a balm for the crisis of spirituality that scientific advances had engendered and which threatened a barren materiality that almost no one wanted. The spectacle of otherwise profound rationalists falling for what seem today to be the most absurd and fanciful of notions – photographs of ectoplasm and fairies, fake mediums got up like spectres, telekinesis and table-rapping – has to be seen in this light. We scoff and laugh at them only if we insist on another fantasy, which is that science is somehow immune to the cultural preconceptions in which it is embedded. Science and technology take on the forms that we need for them. In the late nineteenth century, people needed a spirit world.

It began to seem possible that the Platonists and the medieval theologians were right all along: our visible world was a mere shadow, a feeble projection of a deeper, unseen reality. In the late eighteenth and early nineteenth centuries that notion had been the preserve of mystics and philosophers such as Emanuel Swedenborg, Schopenhauer and Hegel. Now science seemed to agree. 'At an early age', Oliver Lodge wrote in his 1931 autobiography, 'I decided that my main business was with the imponderables – as they were then called – the things that worked secretly and have to be apprehended mentally.' It is not so very different from something Cornelius Agrippa might have said 400 years earlier.

Physicists pushed this dematerialization of the world so hard that the German-Austrian philosopher Johann Bernhard Stallo was moved to complain about it in his 1881 book *The Concepts and Theories of Modern Physics*. The scientific literature, he said,

> teems with theories in the nature of attempts to convert facts into ideas by a process of dwindling or subtilization [that is, making immaterial and invisible]. All such attempts are nugatory; the intangible spectre proves more troublesome in the end than the tangible presence. Faith in spooks . . . is unwisdom in physics no less than in pneumatology.

But he was swimming against the tide. Faith in spooks was what the *Zeitgeist* preferred.

When ghosts turned the tables

In the history of the occult, Spiritualism emerges as a tributary infused with older traditions such as beliefs in ghosts, mesmerism and clairvoyance, while adapted to the nineteenth-century romantic yearning to exceed the constraints of fleeting material existence. In an age when people began to dare hope that they might conquer death, still death came calling, carrying away children, mothers, lovers, whose absence was the harder to bear for being even slightly less inevitable. The nineteenth century has been called an age in permanent mourning; for Lewis Mumford, this was the subconscious reason for the Victorian predilection for black clothing, a fashion endorsed after 1861 by the heartbroken Queen Victoria herself. It was a time when many folk felt a longing for ghosts: they were a source not of terror but of consolation and were regarded no longer as a supernatural phenomenon but a preternatural one, a part of nature that we did not yet understand.

Spiritualism is generally considered to begin with the antics of the Fox sisters Margaret and Kate of Hydesville, New York, who claimed in 1848 to be receiving spirit messages transmitted by rapping noises in their cottage. The Foxes attracted great public interest and by the 1850s they were giving demonstrations – for a fee – of their ability to commune with spirits.

Other mediums began to assert that they too could channel these messages, typically during a sitting or séance in which they gathered with an audience in a darkened room. The rapping spirits of the Fox sisters initially communicated through a crude yes/no code, but this was subsequently refined to pick out messages letter by letter. When Spiritualism reached Europe in the early 1850s, such communications were commonly spelled out instead by table-turning. The attendees at the séance would sit around a table, hands resting on the surface, and it would begin to rotate, tilt or levitate. As the medium called out the alphabet, tilting of the table could identify which letters the spirits selected.

In an earlier time, an occult agency capable of turning tables would have prompted suspicions of demonic activity. But by the nineteenth century, natural philosophers not only accepted invisible forces such as electricity and magnetism but had also begun to harness and explain them. It was natural, then, for those seeking an explanation for table-turning to petition the greatest expert on the invisible forces of electro-magnetism: Michael Faraday at the Royal Institution in London. Several of Faraday's acquaintances urged him to pronounce on these effects and in June 1853 he published a letter in *The Times* reporting the results of careful experiments which set out to detect involuntary movements of the sitters' hands that might explain the rotation of tables. Once an apparatus able to detect a tendency for the sitter to push the table is attached to their hands, said Faraday, 'the power is gone; and this only because the parties are made conscious of what they are really doing mechanically, and so are unable unwittingly to deceive themselves'. Faraday expressed withering dismay at the credulousness of people who attributed the phenomenon to mystical or supernatural agencies. 'I think the system of education that could leave the mental condition of the public body in the state in which this subject has found it must have been greatly deficient in some very important principle', he intoned.

Faraday's studies were utterly convincing and might reasonably have been expected to stifle Spiritualism at birth. That they did nothing of the sort tells us that the mediums were satisfying a widespread desire. They did so with considerable skill and ingenuity, and Spiritualism went from strength to strength. It took on new forms, found scientific supporters of eminence that rivalled Faraday's, and was soon the height of fashion.

The spirit telegraph

It was no coincidence that Spiritualism achieved prominence just as science was demonstrating that communication over long distances was possible. In 1844, four years before the Fox sisters began to attract attention, the American painter-turned-inventor Samuel Morse demonstrated the electric telegraph by sending a message from Washington DC to Baltimore 30 miles away (it said 'What hath God wrought?'). Morse had not invented the device itself; it had been developed in the 1830s by scientists in Germany and in England, where William Cooke and Charles Wheatstone patented a telegraph system that could send messages down a 13-mile cable. But Morse's version was simpler – it required only a single wire for transmission – and with his assistant Alfred Vail he devised the dot-dash alphabet for encoding messages. Morse, already a successful artist, had been motivated to work on a system of long-distance communication after his wife died in 1825 while he was away working on a portrait. By the time the message of her ill health reached him by horse and he had returned to Connecticut to be by her side, she was already buried.

After Morse's official demonstration in 1844, telegraph cables began to snake alongside railways all over the United States, and by the 1850s there were plans to lay them on the sea bed across the Atlantic. The idea of being able to communicate with someone instantly over perhaps hundreds of miles was one that belonged previously to the occult. The German abbot Trithemius of Sponheim had claimed such a power in his book *Steganographia*, written around 1499, in which he said that thoughts might be conveyed over vast distances 'by fire', with the assistance of angels. But it was not just what the telegraph could do that made it seem so wondrous and mysterious, but how it did it. These messages were carried by electricity, at that time still a mysterious force that Mesmer had associated with psychic powers and the Italian physician Luigi Galvani had identified, by setting frogs' amputated limbs twitching using an electrical battery, as the active agent of life itself. (When Morse was seeking funding for his telegraph in the 1830s, they were often discussed by politicians as experiments in mesmerism.) According to Michael Faraday, electricity was a kind of ether that bore the lines of force of an electrical field.

If the telegraph could carry messages by means of this invisible ether, was it so unlikely that some similar disturbance of that occult fluid might mediate between people and spirits, between the living and the dead? Might not departed souls be sending their communications in a Morse code of raps along a spiritual telegraph to the human receivers Kate and Margaret Fox?*

The vast majority of Spiritualist mediums were women. This was easy to understand given the chauvinism of the time: considered passive and lacking in intellect but highly susceptible and impressionable, women were the ideal receivers, halfway between men and machines. Males had too much will and fortitude to become conduits for spirit messages, unless perhaps they possessed an unusually feminine sensitivity. But it was men who tended to investigate these messages, sometimes (as we will see) in ways that crackled with erotic charge. Mediums were literally in the middle, mediators between spirit and (male) intellect: they were, you could say, the first media. Some female mediums were able to exploit this stereotyping to attain a visibility they had previously been denied, and it is not by chance that several of them were aligned with movements for women's rights. At last they had a voice – even if it was not yet permitted to be their own.

Engineers of the ether

One might anticipate that the scientists and inventors who developed telecommunication technologies would have denied that anything miraculous was involved, explaining instead that they were merely manipulating electromagnetic (and acoustic) waves. But the truth was that many of them were also convinced they stood on the threshold

---

* In 1888, five years before she died, Margaret Fox admitted that the raps were produced by cracking the joints of the fingers and feet. 'My sister Katie was the first to observe that by swishing her fingers she could produce certain noises with her knuckles and joints', she wrote in a signed confession, 'and that the same effect could be made with the toes. Finding that we could make raps with our feet – first with one foot and then with both – we practiced until we could do this easily when the room was dark.' Like the Cottingley fairies, this was a prank by pubescent girls that got out of hand. The Fox girls' technique had in fact been revealed in 1851 in a letter from one of their relatives to the *New York Herald* – but what the nascent Spiritualism had to offer was not to be denied by such spoilers.

of more profound insights. They were already deeply immersed in a tradition in which control of invisible forces and emanations was linked to a hidden order of nature, a reality beyond perception which was populated by intelligences and agents. The editor of the *Electrician*, Desmond Fitzgerald, commented ruefully in 1862 that 'telegraphy has been until lately an art occult even to many of the votaries of electrical science.' Alexander Graham Bell, who invented the telephone in the 1870s,* went to séances, his assistant Thomas Watson was a medium in his spare time, and Edison became a member of the American branch of the Society for Psychical Research. 'I am working on the theory that our personality exists after what we call life leaves our present material bodies', he told a reporter in 1920. The telegraph fitted easily into a lineage that includes natural magic, mesmerism, and angels and demons: communication without embodiment, achieved via invisible forces and fluids. Some spoke of Spiritualism as 'celestial telegraphy'.

Cooke and Wheatstone had a hard time convincing Victorian society that telegraphy was a useful and viable technology. To many it looked like just another fancy scientific trick, barely distinct from conjuring: as *The Times* patronizingly put it in 1850, the telegraph served to 'excite wonder at the marvellous feats achieved by modern science'. Scepticism about its practical value seemed to be vindicated when the first attempt to lay a transatlantic cable linking Great Britain to the United States – from the island of Valentia off the Irish coast

---

* Although the credit is generally given to Bell, there were rival claims to his priority – the best known being that of the American inventor Elisha Gray. Another candidate is the Italian inventor Antonio Meucci, a resident of Staten Island, who initially installed an electromagnetic device for transmitting voices in his house to communicate with his wife while she was ill. From the late 1850s to 1870 he made many variants of this 'telectrophone'. Meucci submitted a patent for the device in 1871, and he was hired as an electrician in 1883 by the Globe Telephone Company in New York, which sold his apparatus. But in 1885 the American Bell Telephone Company sued Globe and Meucci for alleged infringement of their patent. Meucci's claim to be the real inventor of the telephone is still a matter of dispute today. But what is seldom remarked is that the Italian inventor's devices were originally inspired by Franz Anton Mesmer's medical use of the 'mesmeric force'. During the patent trial with Bell, Meucci declared that, 'having read the treatise on animal magnetism by Mesmer, I got the idea of applying it by making experiments, using electricity for (treating) ill people, after the suggestion of some physician friends who wished to inquire whether what Mesmer had said was correct'.

to Newfoundland – in 1858 foundered when the cable snapped after 300 miles. This was disastrous for the profile of the industry and helped to stimulate the introduction of a proper scientific training for British engineers. However, with the assistance of William Thomson (who was knighted for his efforts), the cable was finally laid in 1866.

Comparisons with mediumistic transmissions in séances were made explicit by some pioneers of telegraphy, especially Cromwell Fleetwood Varley, chief engineer of the Electric International Telegraph Company. Varley was a bona fide man of science: his mother was related to Faraday, his father was an inventor, and he learnt about electricity at the lectures of William Grove, inventor of the fuel cell, at the London Institution. As a young man he was hired by William Cooke's Electric Telegraph Company, founded in 1845. He was appointed the company's chief electrician in 1858, by which time it had merged with the International Telegraph Company.

For Varley, the phenomena summoned by mediums were no different from any other in nature or the laboratory: they were open to the methods of science. This view was shared by Varley's friend and companion at many séances, William Crookes, who regarded the matter as simply an engineering challenge. As an investigator of spirit communication, Crookes stated that he wished to be considered

> in the position of an electrician at Valentia, examining by means of appropriate testing instruments, certain electrical currents and pulsations passing through the Atlantic cable; independently of their causation, and ignoring whether these phenomena are produced by imperfections in the testing instruments themselves – whether by earth-currents, or by faults in the insulation – or whether they are produced by an intelligent operator at the other end of the line.

Varley became interested in Spiritualism when his wife Ada acquired mediumistic powers in the 1850s. He suspected that spirit messages were entirely analogous to the electromagnetic signals of the telegraph:

> There are probably other powers accompanying electric and magnetic streams, which other powers are seen by spirits, and are by them mistaken for the forces which we call electricity and magnetism.

The challenge for the scientific investigator of Spiritualism, then, was not so different from that which Varley faced in his professional life: to establish a reliable link. According to historian Richard Noakes, he attempted 'to transfer the skills and resources of the telegraph testing-room into a space for spirit communication.'

If telegraphy seemed to confirm the plausibility of Spiritualism, telegraphy *without wires* – radio broadcasting – made that argument even more emphatically. James Clerk Maxwell's equations describing electromagnetic waves predicted that there should exist a continuous spectrum of these waves of all frequencies and wavelengths* humming through the ether. (That word spectrum had once meant 'ghost'.) Vibrations of low frequency, called radio waves, were first generated and detected in 1887 by the German physicist Heinrich Hertz, working in Karlsruhe. Just nine years later the Italian inventor Guglielmo Marconi demonstrated that these waves could be used for long-distance communication, transmitting a message over three kilometres on Salisbury Plain in England. By 1901 Marconi's broadcasts spanned the Atlantic, from Cornwall to Newfoundland. Oliver Lodge (who claimed to have invented wireless telegraphy before Marconi) used these new waves in eye-catching public demonstration experiments. 'Whether vibrations of the ether, longer than those which affect us as light, may not constantly be at work around us', wrote Lodge's friend William Crookes in 1892,

> we have, until lately, never seriously enquired. But the researches of Lodge in England, and of Hertz in Germany give us an almost infinite range of ethereal vibrations or electric rays, from wave-lengths of thousands of miles down to a few feet . . . Here, then, is revealed the bewildering possibility of telegraphy without wires, posts, cables, or any of our present costly appliances.

Radio waves were the first of many invisible rays to be discovered around the turn of the century, and they seemed to offer an entirely

---

* The higher the frequency, the shorter the wavelength. Visible light has frequencies of several hundred trillions of cycles per second (one cycle per second is equal to 1 Hertz) and wavelengths of several hundred millionths of a millimetre. Radio waves oscillate at around thousands to many billions of Hertz, and may have wavelengths ranging from millimetres to hundreds of kilometres.

immaterial telecommunications technology – messages sent invisibly through 'nothing'.

Wireless telegraphy seemed to share the characteristics of paranormal phenomena such as telepathy, a term coined in 1882 by the classicist Frederic H. W. Myers. 'Why, in fact', asked the *Spectator* in 1892, 'if one wire can talk to another without connections, save through ether . . . should not mind talk to mind without any 'wire' at all?' Mark Twain insisted that it was time to take this psychic phenomenon seriously: 'mental telepathy', he wrote in *Harper's Magazine* in 1891, 'is not a jest, but a fact, and . . . is a thing not rare, but exceedingly common.'

But communications borne on the pervasive wireless ether seemed rather different from those channelled down 'spirit wires' to a medium. What was previously a series of directed streams was now a disembodied ocean as vast as the cosmos. And whereas this invisible sea had seemed to Maxwell to be filled with the presence of God, by the *fin de siècle* it had come to seem more like a wasteland for lost souls, broadcasting their electromagnetic lamentation in the hope that someone, somewhere, might tune in – a metaphor, needless to say, for the situation that modernity seemed increasingly to be imposing on the living. The Spiritualist medium had previously acted like a telephone operator, channelling messages to their intended recipient. But now listening for spirits was more like 'DXing', the hobby of the first radio hams: scanning the airwaves for random signals from distant stations. These broadcasts were omnipresent, not only unseen but also unheard unless you happened to tune in to the right frequency. Like spooks, they enter the home unbidden – you couldn't shut them out even if you wanted to. As *Scribner's Magazine* put it in 1923 in an article on radio communication called 'Eavesdropping on the world',

> We have been toying with the intangible, the eerie something out of which Northern Lights are made, the ripples in the boundless vastness of space. Who knows where it will lead us as we bend the mysterious forces to our call and read the thoughts flying on the wings of the ether?

While Victorian séances had promised consolation, there was now a sense of foreboding, of weird apparitions and threatening creatures that might emerge from the invisible etheric depths. Radio waves were rumoured to be picked up by inanimate objects to give them a bizarre semblance of life: a shovel that sang in Sweden, a metal drinking fountain on a farm that hummed a symphony, a mirror that, as if in a fairy tale, spoke to a woman one morning as she gazed at her reflection. 'As popular knowledge of radio principles grew', writes cultural historian Jeffrey Sconce,

> Americans began to realize that they were all continually negotiating an invisible world of radio waves, whether they wanted to be or not, and that anything from a coal shovel to shoddy dental work could serve as a potential gateway into the mysterious realm of ether.

Rudyard Kipling captured this perception of random occult communion with spirits in his short story 'Wireless' (1902). On a freezing winter night, the narrator arrives at a chemist's shop on the south coast of England to witness an amateur experiment with Marconi's new device and the 'Hertzian waves' – an attempt to send a message to a receiver station many miles away in Poole, Dorset. The chemist's assistant lies ill with consumption, and in a drugged trance he 'picks up' and writes down lines from a poem by Keats, who he has never read. At the same time the chemist's nephew hears similarly disjointed fragments of speech on the wireless – meaningless snippets from ships offshore. 'Have you ever seen a spiritualistic séance?' he says. 'It reminds me of that sometimes – odds and ends of messages coming out of nowhere.'

Kipling apparently believed that minds might receive random thoughts vibrating through the spiritual ether. 'We are only telephone wires', he wrote to Rider Haggard in 1918. But even telephone wires were no longer needed to send the message; all it took was for the brain to act like a transmitter and receiver of unseen, information-laden rays. J. J. Thomson, head of the Cavendish Laboratory in Cambridge, endorsed experiments to test whether telepathy could be caused by the leakage of cranial electromagnetic signals. If Thomson felt that to be possible, who could disagree?

Crookes's ghosts

Varley and Crookes weren't uncritical dupes – they acknowledged that Spiritualism was full of charlatans and frauds. But that's not all it was. Many mediums genuinely believed they were in contact with the spirit world. Perplexingly, some of them found it possible to sustain that belief *even while* perpetrating what we would now regard as hoaxes – just as, perhaps, some scientific fraudsters today apparently believe their faked results to be only a shortcut to honest truth. This ambiguous character of Spiritualism is particularly well attested by the investigations that Crookes made into its invisible realms.

William Crookes was a self-made man, the son of a London tailor who invented himself in many guises: journalist, industrial analyst, consultant and independent chemist – deriving money and influence from a diverse portfolio of interests ranging from scientific publishing to public sanitation and gold mining. While Crookes's older step-brothers took up their father's profession, William showed an early interest in chemistry, inspired by the lectures and spectacular demonstrations he attended at the Royal Polytechnical Institution near his home. John Henry Pepper, later the impresario of the theatrical ghost, was not appointed there until 1848, in which year the 16-year-old

Sir William Crookes (1832–1919), chemical entrepreneur, publisher and Spiritualist.

Crookes enrolled at the Royal College of Chemistry in Oxford Street. But Crookes would have got to know this flamboyant lecturer during his time at the college, for the Chemical Society of London, of which Crookes was a young member, took to meeting at Pepper's house.

Crookes applied his chemical passions to the new art of photography. At the Royal College of Chemistry he devoted much of his spare time to taking photographs and investigating how the techniques might be improved. He gained a reputation as something of a youthful authority on these matters, and began corresponding with Henry Fox Talbot, who called on Crookes as an expert witness in a patent dispute in 1854. Three years later Crookes became editor of the *Liverpool Photographic Journal* (which dropped the provincial 'Liverpool' in 1859) – the first of his forays into scientific and technical publishing, which he followed by launching *Photographic News* in 1858.*

Crookes was constantly juggling interests in science, business and publishing while looking out for marketing opportunities or ways to convert his knowledge into hard currency. Sifting through other journals and periodicals for material to fill the photographic magazines left him with surplus that he decided to collate in a general chemistry magazine. He called it *Chemical News*. Launched in late 1859, it became a platform for Crookes to indulge and advertise his voracious appetite for all things chemical, from mining to cattle plague. It also betrayed his somewhat undiscerning attitude: he would print just about everything that was submitted, 'leaving the professional reader and time to sort the wheat from the chaff', in the words of chemistry historian William Brock.

Crookes constructed a private laboratory in his London house in Mornington Road and the results of his researches often went straight onto the pages of *Chemical News*. Yet he was no dilettante and in 1861 he announced the discovery of a new chemical element. Two years earlier the German scientists Robert Bunsen and Gustav Kirchhoff reported that the light emitted when chemical elements burn in a flame contains a fingerprint of their identity. When a prism is used to split this light into

* We saw in the previous chapter how photography fed into the 'scientification' of Spiritualism. The American brothers William and Ira Davenport, whose touring show of illusions in the 1850s and '60s was presented as a genuinely supernatural spectacle (until their fakery was exposed by P. T. Barnum in 1865), argued that they needed to perform in near-darkness because, like photography, the results depended on it. Which in a sense they did.

its component wavelengths, one finds bright 'spectral lines' at particular wavelengths characteristic of each element. Using this technique to analyse the sludgy residue left over from the production of sulphuric acid, Crookes discovered a line of green light not associated with any known element and presumably therefore due to a new one. He called it thallium, after the Greek for 'new twig', alluding to the leaf-green hue of its spectral emission line. Crookes's evidence for this new element was tentative at first, and the announcement was arguably premature (as well as being concerned to secure priority, he was apparently short of material for that issue of his journal). But when the French chemist Claude-Auguste Lamy isolated this metallic substance the following year, the reality of thallium was confirmed, and with it Crookes's position in British science.

It was, however, a curious position. Lacking a university education and an academic position, blessed with little cultural sophistication or personal charisma, Crookes was respectable without ever quite belonging to the establishment. When he began to display sympathy for Spiritualism, he was therefore all the more vulnerable to criticism, even if a passing interest in such matters was not at all unusual for nineteenth-century scientists.

Crookes's interest in the spirit world seems to have been triggered by the death of his beloved younger brother Philip from yellow fever in the 1860s, ironically during an expedition to Cuba to lay a telegraph cable. Crookes began to attend séances, and in 1868 he became persuaded that the 18-year-old medium Frank Herne was channelling his dead brother's messages. In all likelihood, Herne (or his manager – most mediums possessed one) had researched the backgrounds of his guests and come across news of the libel suit between Crookes and the captain of the ship to which Philip had been assigned, whom Crookes had accused of negligence.

Crookes became active in Spiritualist circles and attended séances by Cromwell Varley's wife Ada. In 1869 he wrote to John Tyndall, a sceptic of all things supernatural and an advocate for the total divorce of science from religion, saying that

> Something new and worthy of the notice of the man of science I am tolerably certain we are getting glimpses of, and I fancy my thoughts are shaping in the direction of a power in some way connected with gravitation.

He became convinced that the most famous medium in Britain, Daniel Home, could modify gravity by manipulating a 'psychic force'. Here he was, then, stitching back together the old web of occult forces, which could be tugged this way and that by those who knew how.

In that age of charlatans, few were more accomplished or more slippery than Home. Born in Edinburgh, he lived with his aunt in Connecticut until coming to London aged 22 to seek a lucrative audience for his trickery. For the next decade he wandered the courts and salons of Europe and was banished from Rome as a sorcerer on command of the Pope. In 1868 he was forced by the English law courts to relinquish the fortune and estate he had swindled from a credulous widow. By 1869, when Crookes met Home, there was already ample reason to regard him as a fraud. But he was charming, and Crookes was captivated.

Home put on dazzling, even dangerous displays. Here is Crookes's account of events during a séance in May 1871, as the participants sat around the table in darkness with hands linked:

> At first we had rough manifestations, chairs knocked about, the table floated 6 inches from the ground and then dashed down, loud and unpleasant noises bawling in our ears and altogether phenomena of a low class. After a time it was suggested that we should sing, and as the only thing known to all the company, we struck up 'For he's a jolly good fellow'. The chairs, tables and things on it kept up a sort of anvil accompaniment to this. After that D. D. Home gave us a solo – rather a sacred piece – and almost before a dozen words were uttered Mr Herne was carried right up, floated across the table and dropped with a crash of pictures and ornaments at the other end of the room. My brother Walter, who was holding one hand, stuck to him as long as he could, but he says Herne was dragged out of his hand as he went across the table . . .

The group was subsequently treated to accordions playing themselves, floating lights, books dashed about and disembodied hands stroking their faces. The effect must surely have been overwhelming – both exciting and frightening, and doubtless calculated to inhibit objective assessment. Accounts like this – and there are many of them – attest to the energetic and ingenious preparations that must have gone into séances.

Between 1870 and 1873 Crookes, assisted by Varley, conducted over 30 sittings with Home, during which they subjected him to all manner of 'scientific' tests. Although Crookes admitted that the medium 'is himself subject to unaccountable ebbs and flows of the force, [and] it has seldom happened that a result obtained on one occasion could be subsequently confirmed and tested with apparatus specially contrived for the purpose', nonetheless he was convinced that Home was no phoney. Even as late as 1889 he insisted that 'certain of Home's phenomena fall quite outside the category of marvels producible by sleight of hand of prepared apparatus.'

What, then, *was* their cause? Crookes was sure that some unseen, purposeful agency was at work. In 1874 he wrote to one of Home's disciples, a Madame Boydanof in St Petersburg, saying '[a]ll I am satisfied of is that there exist invisible intelligent beings, who *profess* to be spirits of deceased people, but the proofs which I require I have never yet had'. As a scientist he felt that his role was to investigate the modes by which such beings communicate with us: to improve these channels just as he and others sought to improve the telegraph. To the medium Jane Douglas he wrote in 1871 saying that

Assuming that there are invisible beings trying to communicate with us, it is reasonable to suppose that improvements can be made in their mode of telegraphy, and whilst others are obtaining copiously worded communications, I prefer to devote myself to the humbler but not less useful work of acting as a telegraphic engineer, endeavouring to improve the instrumental means at this end of the line, to ascertain conditions which will render intercourse more certain, and generally to get the line in a good state of insulation.

Few statements put the situation in plainer perspective, juxtaposing as it does the most fantastical speculations ('Assuming that there are . . .') with the notion that an engineer might twiddle the dials and improve the reception from this netherworld in much the same way as he could improve the radio signal from Helsinki. True to his word, Crookes set about developing what might now be regarded as a kind of occult technology.

Home agreed to come to Crookes's home laboratory to be subjected to scientific testing, and in 1871 Crookes announced that he was searching for the psychic force by which Home achieved his astonishing feats. His experimental methods left something to be desired. One test involved Home playing a melody on a concertina that was surrounded by a wire cage and placed under the table at which he sat, while one or both of Home's hands were prevented from touching the instrument. In another experiment, a thick wooden plank was placed on a fulcrum with a spring balance at one end and Home was able to increase its apparent mass by about four kilograms just by touching the other end with his finger. Both are tricks known to illusionists. A metal bracket and body harness hidden by a shirt can impart inordinate strength to a single finger, while Victorian music-hall artists would appear to 'play' accordions remotely using a tiny harmonica hidden inside the mouth, further concealed perhaps behind exuberant Victorian whiskers. Crookes should of course have checked for both possibilities by thoroughly searching his subject, but to do so would have revealed an ungentlemanly lack of trust. Behind closed blinds, it was not hard to deceive a man with Crookes's poor eyesight.

In June 1871 Crookes sent a paper describing his findings to the Royal Society, which rejected it on grounds of poor experimental method. So Crookes published it in his own scientific periodical, the *Quarterly Journal of Science*. 'These experiments', he wrote,

> *confirm beyond doubt* the conclusions I arrived at in my former paper, namely, the existence of a force associated, in some manner not yet explained, with the human organization, by which force increased weight is capable of being imparted to solid bodies without physical contact.

He reported that this force was not at all reliable, unlike gravity or magnetism, but could change in strength from one hour to another, sometimes disappearing altogether and then revitalizing 'in great strength'. He claimed that it could act up to 2–3 feet away from Home but was at its strongest near to him. At this time Crookes was studying anomalies in the apparent atomic weight of thallium, which he thought might be explained by a force that opposes gravity. He began to suspect that Home's psychic force too might be a kind of anti-gravity, mediated by electromagnetic radiation. In the thallium studies the weight seemed sensitive to differences of heat in the samples, and so Crookes was excited by the idea that some link between gravity and heat was being exposed that the psychic force might elucidate.

Light mills and dark spaces

Crookes devised instruments to measure these subtle, invisible forces. With his assistant Charles Gimingham he made an extremely sensitive torsion balance, consisting of two pith balls connected by a straw or a glass rod and suspended from a silk thread, all encased within an evacuated glass vessel. Tiny forces experienced by the balls could make the armature turn – although of course so might all manner of very slight vibrations and other disturbances, making the experiments hard to interpret. Crookes and Gimingham later refined the design, giving the rotor four square metal vanes attached to a central pivot.

Crookes found that when sunlight was shone onto the vanes, the arm turned as though they were being repelled by the light, suggesting that light exerts a kind of pressure. Might this 'radiant molecular energy' be the origin of gravity, he wondered? The effect was stronger if one side of the vanes was painted black. Crookes anticipated that the white or reflective metal faces of the vanes should experience the greater 'aetherial pressure', since more light bounces off them whereas the black faces absorb it. But in fact the rotation is greatest when light falls on the *black* sides.

One of the 'light mills' or radiometers made by William Crookes and Charles Gimingham to detect very weak forces, including alleged 'psychic forces'.

This puzzle notwithstanding, when James Clerk Maxwell was sent Crookes's 1875 paper on the 'light mill' to review, he recommended that it be published, for it seemed to verify his prediction two years earlier (of which Crookes had been unaware) that light should create mechanical pressure. The speed at which the arms rotate was found to depend on the intensity of the incident light, for which reason the light mill is also known as a radiometer: an instrument for measuring light. Crookes estimated that sunshine falling on the earth exerts a pressure of 2.3 tons over every square mile, and in 1876 he gave a talk at the Royal Institution on 'weighing a beam of light'. His invention was widely praised – one journal said of it, with somewhat florid hyperbole, that 'science has made a hole in the infinite' – and the Royal Society made amends for their treatment of Crookes's psychic research by awarding him a Royal Medal in late 1875. After the German astrophysicist Karl Zöllner visited Crookes that year, he returned home to ask the experimental physicist Heinrich Geissler, inventor of the gas discharge tube in which cathode rays were discovered, to make him a radiometer too.* Soon the enterprising Geissler was supplying them all over Europe.

It turns out, however, that the vanes rotate not because of light pressure (although this is in itself real enough) but because of the impact of gas molecules within the imperfect vacuum of the light mill's glass chamber. But the reason is subtle. The black sides of the vanes absorb more of the light energy than the white, and so they are warmer and will make the nearby gas warmer too. In 1879 Osborne Reynolds, a physicist at the University of Manchester, discovered that gas molecules will pass preferentially from cool to hot across such a temperature difference: an effect called thermal transpiration. Some gas molecules collide with the edges of the vanes as they pass by, imparting a little kick to the vanes. As more molecules pass from the warm (black) side to the cool (white) side than vice versa, overall the vanes get a little push in this direction. James Clerk Maxwell understood all this when he saw Reynold's work, and he published the correct explanation shortly before his death in 1879. (Reynold's paper was published later than Maxwell's, so only belatedly did he receive due credit.) The effect also explains why Crookes found that cold objects held close to the vanes

---

* Whether Crookes played any part in it or not, Zöllner was later converted to Spiritualism and spent the rest of his career developing a 'transcendental physics'.

seem to attract them (Crookes himself interpreted this as further evidence of a connection between heat and gravity).

In this way, Crookes's Spiritualist inclinations motivated him to undertake important scientific experimentation. As Crookes's biographer William Brock says, '[w]hile Crookes, arguably, would still have developed the radiometer without the influence of Spiritualism, it provided him the personal spur to enter the new world of microphysics' – and to make an iconic piece of equipment. 'When the radiometer was first constructed in the mid-1870s', says Brock, 'it must have seemed an extraordinary and mysterious object and one that might, indeed, link the worlds of material scientific reality with the mysterious and religious unknown.' Philip Pullman alludes to the device and its associations in the *His Dark Materials* trilogy, in which the revolutions of a radiometer in his 'alternative Oxford University' take on mystical relevance:

> At Gabriel College there was a very holy object on the high altar of the Oratory, covered . . . with a black velvet cloth . . . At the height of the invocation the Intercessor lifted the cloth to reveal in the dimness a glass dome inside which there was something too distant to see, until he pulled a string attached to a shutter above, letting a ray of sunlight through to strike the dome exactly. Then it became clear: a little thing like a weathervane, with four sails black on one side and white on the other, began to whirl around as the light struck it. It illustrated a moral lesson, the Intercessor explained, for the black of ignorance fled from the light, whereas the wisdom of white rushed to embrace it.

The conscious, semi-spiritual cosmic particles known as 'Dust' in Pullman's books were in turn inspired by a new, radiant state of matter that Crookes believed his studies with the radiometer had revealed, and which he interpreted as a possible link between the visible and invisible worlds. 'In studying this fourth state of matter', said Crookes,

> we seem at length to have within our grasp and obedient to our control the little invisible particles which with good warrant are supposed to constitute the physical basis of the universe.

What was this 'radiant matter'? In one sense it amounted to nothing more than the residual gas remaining in the radiometer's chamber after

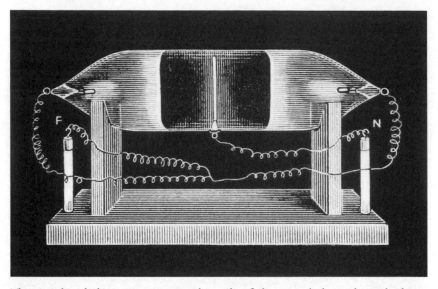

The Crookes dark space, seen on either side of the central electrode (cathode) in this Geissler discharge tube. Crookes described the apparatus: 'Here is a tube, having a pole in the center in the form of a metal disk, and other poles at each end. The centre pole is made negative, and the two end poles connected together are made the positive terminal. The dark space will be in the centre. When the exhaustion [vacuum] is not very great, the dark space extends only a little on each side of the negative pole in the centre. When the exhaustion is good, as in the tube before you, and I turn on the coil, the dark space is seen to extend for about an inch on each side of the pole.' Faraday's dark space is not visible in this experiment.

it was evacuated. But ultimately it presaged one of the pivotal discoveries of late Victorian physics. Crookes claimed that the tenuous gas inside the chamber had properties 'as far removed from those of a gas as this is from a liquid'. He felt that a vital clue to its nature was indicated by the 'dark space' that could be seen close to the negative electrode (cathode) when an electrical discharge passes through a very low-pressure gas in a Geissler tube. These glass tubes, in which an electrical field converted some of the gas molecules to electrically charged ions which then glowed as they conducted an electrical current between the electrodes, were central to experimental investigations into the connections between light, electricity and matter in the nineteenth century.

Crookes stole the notion of radiant matter from Michael Faraday, as he later admitted. In 1816 Faraday – still then a student – had speculated about the existence of some rarefied substance at the boundary between light and ordinary matter:

If we conceive a change as far beyond vaporization as that is above fluidity, and then take into account also the proportional increased extent of alteration as the changes rise, we shall perhaps, if we can form any conception at all, not fall far short of radiant matter; and as in the last conversion many qualities were lost, so here also many more would disappear.

Crookes's 'dark space' was an appropriation from the same source, for Faraday had reported something similar close to the cathode of a Geissler tube. But Crookes was looking at a slightly different region, immediately surrounding the cathode. This came to be called the 'Crookes dark space', as distinct from the 'Faraday dark space' further from the cathode.

Crookes came to suspect that his dark space was an emanation of the putative 'fourth form of matter'. He found that, under the right conditions, he could extend the dark space into a kind of beam projecting out from the black vane of a radiometer when it was wired up to act as the cathode of a discharge tube. Where this dark beam intersected the wall of the glass vessel, it created heat along with a greenish phosphorescent glow. In other words, he said, this dark space was like a beam of 'molecular light' or 'radiant matter', generating visible light when it hit the glass. The glow could be deflected by a magnet, which was not the case for ordinary light. 'The phenomena in these exhausted [that is, vacuum] tubes', Crookes wrote,

> reveal to physical science a new world – a world where matter exists in a fourth state, where the corpuscular [particle] theory of light may be true, and where light does not always move in straight lines; but where we can never enter, and in which we must be content to observe and experiment from the outside.

In some public talks Crookes stepped over the precarious line he had created by suggesting that radiant matter was the link between the physical and the occult worlds. But after all, why not? This was physics as strange and unearthly as anything one can find in Pullman's novels. At a talk on radiant matter at the British Association meeting in 1879 Crookes said:

We have actually touched the border-land where matter and force seem to merge into one another, the shadowy realm between known and unknown, which for me has always had peculiar temptations. I venture to think that the greatest scientific problems of the future will find their solution in this border-land, and even beyond; here, it seems to me, lie ultimate realities, subtle, far-reaching, wonderful.

These studies weren't entirely original. In 1858 the German physicist Julius Plücker had reported 'rays' emanating from the cathode of a discharge tube, showing that they could make the glass wall glow and that they could be deflected with a magnet. Plücker's colleague Johann Wilhelm Hittorf showed that the rays travelled in straight lines, since the glow would be extinguished if a solid object were placed between the cathode and the tube wall. In 1876 the German physicist Eugen Goldstein called them 'cathode rays', and it was assumed by researchers in Germany that they were a form of electromagnetic radiation. But Crookes, now appropriating cathode rays to himself, argued that they were particles of 'radiant matter', which carried electric charge. In apparent support of that idea, he and Gimingham showed that a tiny paddle made of a flake of mica could be pushed along a rail when struck by a beam of cathode rays.*

Crookes came up with a better way to register the impact of these invisible rays/particles. Instead of relying on the phosphorescence of the glass tube, he used magnets to direct the cathode-ray beam onto a phosphor: a screen of mica laced with the mineral zinc sulphide, which glowed when the rays hit. This became known as the Crookes tube, or also simply as the cathode-ray tube, and it eventually morphed into the television, the vehicle that allowed glowing dreams to be plucked out of the invisible ether and brought into the domestic hearth. For as J. J. Thomson demonstrated in 1895, cathode rays were in fact the negatively charged particles called electrons: not exactly a new state of matter, but the first intimation that atoms have more fundamental component parts.†

---

* In fact the paddle, like the radiometer's vanes, moves because of heating effects, as J. J. Thomson showed. But, to spoil the punchline that follows, it is still sometimes said that this experiment by Crookes demonstrates for the first time that electrons (for it is they that underpin the dark space and its light-eliciting nature) are particles.
† Thomson explained that the mass of the electron was the result of its electrical

As we will see, work on cathode rays soon led to the discovery of X-rays and radioactivity. Because he used phosphors to reveal them, Crookes befriended the French expert on phosphorescence Alexandre-Edmond Becquerel, whose son Henri discovered the 'uranic rays' emanating from uranium that the Curies christened radioactivity. These rays heralded a century of new extremes of light and dark, brighter than a thousand suns and stygian as the world's end. Half a century later and on the other side of the world, they were destined to cast shadows burnt onto municipal stonework like the imprints on photographic plates, while the people whose shapes they recorded had, like their city, vanished.

The fifth force

Crookes's attempts to use exquisite scientific devices to find a psychic force may look odd today, but delicate rotating instruments that might detect occult influences were all the rage in the late nineteenth century. Other instruments to measure the subtle effluvia of human bodies included the magnetometer of Abbé Fortin, the biometer of Louis Lucas (a kind of galvanometer), the bioscope of Hippolyte Baraduc, the sthenometer of Paul Joire, and the 'fluid motors' of the Comte de Tromelin, the latter consisting of paper cylinders on needle points that rotated when a human hand was near. All of these individuals were French occultists or parapsychologists, several of them influenced by Eliphas Lévi. Many were inspired by the discoveries of X-rays and radioactivity, which seemed to suggest that there might be yet more invisible emanations from the human body that could be detected with the right apparatus. The French physicist Jules Thore claimed to have found such a subtle force in the 1880s, using a cylinder of ivory suspended by a silk thread which would rotate if a second cylinder was brought to within about a millimetre. Crookes repeated the experiments in 1887 and found that the results depend on the heat of the human body.

---

interactions with all the other electrons in the universe, which exerts an inertial drag – a strikingly similar argument to the way the mass of other fundamental particles is now regarded to be a consequence of their interactions with the much celebrated Higgs particle. But Thomson spoke of the electron thereby acquiring an 'etherial or astral body' – a remarkable example of the importation of explicitly occult imagery into this formative atomic science.

A torsion balance used by Eötvös' assistant and successor Dezsö Pekár in the 1930s to make measurements of variations in the earth's gravity. The armature rotates around the vertical axis.

Again we shouldn't mock. Much the same kind of instrumentation – the suspended rotating needle or bar – is still used today to search for a scarcely less speculative hypothetical force that might operate between material bodies. In the late nineteenth century the Hungarian physicist Lórand Eötvös used a torsion balance – a bar suspended from a wire and bearing a heavy mass at each end – to look for tiny changes in the force of gravity from one place to another at the surface of the earth owing to differences in density of the underlying rock. Eötvös devised the apparatus initially to study the supposed equivalence of two definitions of a body's mass: its resistance to being moved by an applied force (its inertial mass) and its attraction by the force of gravity (its gravitational mass). The so-called 'weak equivalence principle' asserts that these masses are identical; it was a central component of Einstein's theory of general relativity in 1915, which explained gravity as a distortion of spacetime.

In 1986, US physicists reanalysed the measurements Eötvös and his colleagues had made, and found that they seemed to indicate that an object's gravitational mass depends not only on the combined mass of all its atoms but on their chemical nature: different materials of the same notional mass have a different gravitational mass. In other words, the researchers said, there is a 'fifth force' (that is, one additional to the four currently known fundamental forces of nature: electro-magnetism, gravity and the two nuclear forces) that seems to make gravity dependent on the material on which it acts. If it exists at all, this force must be extremely weak: about the same strength as gravity itself, which is minuscule unless the objects are at least the size of mountains. Several modern experiments using even more sensitive torsion balances and other instruments have sought to confirm this 'fifth force', but none has conclusively done so. Theoretical justifica-tions for a fifth force have, however, been found in some versions of string theory, which invoke extra dimensions of space beyond the three that we experience (see Chapter 5): another invisible force emanating from an invisible realm. The fifth force may well prove to be a chimera. That this unproven idea is sanctioned at all in modern physics testifies to the tenacious appeal of subtle, invisible and perva-sive forces of a kind we can as yet barely imagine. Crookes was no less modern for believing in them.

Ghost clubs

Crookes had to curtail his investigations involving Home when the medium moved to Paris in October 1871. By this time Crookes had realized that studying 'psychic' phenomena would create trouble in scientific circles. It didn't deter him, but made him rather more clan-destine about such activities. In 1872 he found a new subject: the 16-year-old Florence Cook, who lived with her family in Dalston, northeast London. Like many young female mediums, Cook was striking – if not conventionally beautiful, nevertheless of so commanding an appearance that one must wonder about her effect on an elderly gentleman like Crookes. Alfred Russell Wallace, another eminent Victorian scientist swayed by Spiritualism, was also much taken by Cook. During a séance she would retire from the room into a curtained alcove ('cabinet') and fall into a trance, and in her place appeared a spirit called Katie King. King would not, needless to say, look exactly like Cook – she was allegedly taller, with different coloured hair and skin – but she seemed strangely substantial for a phantom.

The Victorian medium Florence Cook (*c*.1856–1904).

On some occasions Crookes walked arm in arm with her around the room and 'asked her permission to clasp her in my arms' – whereupon he found her to be 'as material a being as Miss Cook herself'.

Might it not have occurred to Crookes that Miss King *was* Miss Cook? He denied this possibility, emphasizing their differences in appearance, but that conclusion is plain enough from the account of one of Cook's séances in 1873 by the politician Lord Arthur Russell:

> I had been led by the accounts of witnesses to expect a startling appa-
> rition; it was therefore, naturally, very disappointing, after Miss Florence
> Cook had been tied down in the cupboard, and the ghost of 'Katie'
> looked out of the peephole, to observe that the face of the ghost was
> merely Miss Florence Cook's face, with a piece of white linen wrapped
> around it, and that the black face that subsequently appeared was again
> merely Miss Cook's face with a black tissue drawn over it.

It is possible that Cook sometimes employed an assistant, however (perhaps her younger sister Kate), for occasionally spirit and medium were 'seen' together – although never so that both their faces were visible. Crookes even took photographs of the two of them, but one or other always had her face obscured.

The improbability that Crookes, a skilled photographer, would have failed repeatedly to capture both visages by accident is one reason why he has been accused of being Cook's willing accomplice in the deception. One commentator, Trevor Hall, suggested in 1962 that Crookes was drawn into collusion because of an affair with Cook – rumour to that effect inevitably circulated in Crookes's time. William Brock doubts this; the best one can say is that the evidence does not exclude it. But whether or not their liaison was sexual is rather beside the point (beyond the obvious scandal it would have caused), for there seems no doubt that Crookes was infatuated with Cook/King to a degree that compromised at least his objectivity, if not his integrity. His gushing words in *The Spiritualist* magazine in 1874 might just as well be those of a clandestine lover: 'Photography is as inadequate to depict the perfect beauty of Katie's face, as words are powerless to describe her charms of manner.' He reverently cut locks from Katie's golden hair (so unlike Cook's dark tresses!), put his ear to her bosom to hear her heart beat and watched her sit and tell his children stories of her exploits in India. (King claimed to be the spirit of the daughter of the seventeenth-century buccaneer Henry Morgan, now immortal-ized on bottles of his eponymous rum.)

Crookes himself mentions that he was so far taken into Cook's confidence that she would permit him to sit with her in the cabinet while Katie King appeared before the other sitters. Quite aside from the physical intimacy this implies, it forces one to ask how on earth Crookes could have been anything other than deceitful about Cook's costumed manifestations, unless on those occasions she did indeed use an accomplice. Certainly Crookes was permitted to assume almost a managerial role in the sittings, leading Hall to suggest that he 'was thus the impresario who chose the sitters, controlled the séances and was allowed inside the cabinet.'

One of the supposedly most demanding 'scientific' tests devised by Crookes and Varley involved making the medium an element in an electrical circuit, so that, were she to have vacated the cabinet to pose as the spirit, her absence could be recorded and detected as a break in the current flowing through the circuit. To this end, the medium would be asked to sit with brine-soaked blotting paper attached as electrical contacts to her body, or grasping two electrodes. Varley and Crookes created instruments that used a moving pen to record a trace of the electrical resistance of the circuit. Here is perhaps the most striking representation of female mediums as passive machines: they have become mere components of electrical circuitry. Of course, one might instead have simply kept the medium under close observation to ensure she did not leave her place – but that would have been considered a breach of etiquette.

Cook apparently passed the circuit-breaker test several times. One suggestion is that, as it took a few seconds for an interruption in the circuit to register on the recording device, she might have quickly faked her presence by slipping the loops and placing an electrical resistor between them. This would seem to require that she knew exactly what the electrical resistance of her body was, which has fuelled suggestions that Crookes actively collaborated in the deception. In any event, the apparatus served again to create a conceptual link between the electrical circuits of the telegraph and the channels of Spiritualism.

Cook was most definitively exposed as a fraud in 1880, when news-papers reported how a sitter had discovered her change of clothes. (That didn't prevent her from continuing to give séances almost up until her death in 1904.) There were plenty of other exposés too that ought to have dampened Crookes's enthusiasm. In 1875, days after Cook performed a séance in his laboratory with another young medium, Mary Rosina Showers, who manifested as a spirit called 'Florence', the two young women were invited for a repeat perfor-mance at the home of Crookes's friend, the lawyer Serjeant Edward Cox. Their spirits were to appear from behind a curtained-off part of the room concealing the mediums lying in a trance. But Cox's young daughter, oblivious (so Cox said) to the conventions of these gather-ings, rose from the table and pulled back the curtain after 'Florence' had just disappeared behind it, to reveal Showers in the spirit's head-dress but half-changed back into her black gown.

Showers confessed her fakery that year to the American medium Annie Fay, another of Crookes's subjects, who duly told him. Crookes confronted Showers and declared that he would expose her if she did not desist from the fraud. But he was no match for the quick-witted young woman, who secured a promise from him that he would not tell her mother about the business. So when said mother decided that

Crookes's dealings with her daughter had a sexual aspect, he was honour-bound not to tell her the truth of the matter. The gullible Crookes was hoodwinked by Fay too; she privately confessed in 1913, although a keener observer might have expected as much, for Fay never really distinguished herself from any other Victorian illusionist, advertising her performance at the Queen's Hall in London in 1874 as a 'series of bewildering effects' and 'mysterious manifestations'. Fay's career reminds us that the mediums sometimes occupied the same intentionally ambiguous middle ground as stage magicians, cultivating an air of wonder without explicitly stating whether they were performing illusions or communing with spirits.

Despite all this, Crookes defended the claims of Spiritualists passionately in the face of his scientific colleagues' scepticism. When his friend John Spiller, a photographic chemist, wrote in the *Spectator* about the exposure of Daniel Home's deceptions, Crookes ceased to communicate with him for a quarter of a century. He responded to the doubters by pointing out that they too relied on 'invisible' mechanisms. Didn't physicists and chemists invoke unseen atoms? 'Yet these ultimate particles, molecules or atoms are creatures of the imagination and as pure assumptions as the spirits of the Spiritualists,' he declared.

It isn't clear whether Crookes was simply a fool or something more corrupt. But his mistake was not that he took Spiritualism and the paranormal seriously, but rather that he was so little able to apply scholarly standards to their investigation. Perhaps he was indeed a fraud himself. But the judgement of modern 'fraud-busting' magician James Randi is probably closer to the truth:

> He believed, in common with so many other dupes, that he was far too smart to be fooled; in actuality, he was not smart enough to recognize that he was the perfect patsy.

Invisible chemistry

In 1883 Crookes joined the Theosophical Society after being invited by the leader of the London branch, Alfred Percy Sinnett. Four years later he allied himself with the Esoteric Section of the London lodge, founded by Madame Blavatsky, who was then locked in a power struggle with Sinnett. Blavatsky recognized the value of scientific supporters such as Crookes, and she flattered his ego by referring to his work on radiant matter in her *Secret Doctrine* (1888), the foundational text of Theosophy, which cleverly exploited scientific disagreements about invisible concepts such as force, light, atoms and the ether to assert the case for considering her occult alternatives.

Of the strange progeny of Theosophy's marriage of modern science and mysticism, few were odder than *Occult Chemistry*, a book published in 1908 by Charles Leadbeater and Annie Besant. Leadbeater was the archetypal Edwardian libertine: sporting purple cloaks, infatuated with India and preaching sexual liberation that leaked into rumours of pederasty. His 1902 book *Man Visible and Invisible* was full of rather crude paintings of the 'astral bodies' of different character types, replete with the conventional racism of the age in which the aura of the 'savage' is marked by selfishness, deceit, greed and sensuality (although some savages 'would compare favourably with the lower specimens of our own civilization'). Besant was another matter. A formidable personality, devoutly religious, beautiful and independent (she separated from an unhappy marriage in 1873), she was a fearless social campaigner, arguing for women's rights and birth control, better conditions for workers, free school meals, educational reform and home rule for India. Her conversion to Blavatsky's pseudo-religion in 1889 dismayed many of her progressive friends, including George Bernard Shaw, but it never seemed anything less than heartfelt.

In *Occult Chemistry* this somewhat unlikely duo described how, using meditation techniques suggested by Sinnett to shrink their perception to microscopic scales, they were able to 'see atoms, using etheric eyes.' They claimed to have identified several new chemical elements this way, including one they called Occultum. Most of the book was taken up with elaborate drawings of the shapes of atoms of different elements: some like ornate frozen splashes, others resembling the rose windows of Gothic cathedrals. Far from being uncuttable – this was eight years before Ernest Rutherford astonished the world by 'splitting the atom' – these atoms are composites of more fundamental particles called *anu*, the Sanskrit word for atoms. Leadbeater and Besant's hydrogen is composed of 18 anu and nitrogen contains no fewer than

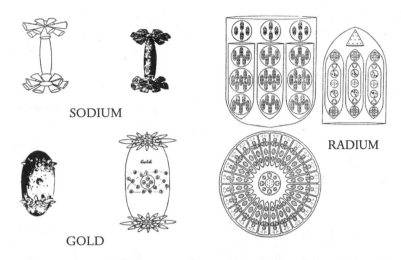

SODIUM

GOLD

RADIUM

Atoms of gold, sodium and radium as 'visualized' by Leadbeater and Besant in *Occult Chemistry* (1908).

290 of them. The anu are themselves 'atoms of ether', which is imagined here as a particulate medium.

Crookes was already well disposed to the idea that atoms were composite entities – he had suggested that they were made of particles of a fundamental substance called protyle. Besant evidently regarded Crookes as a potential ally, and sent him a preliminary account of her 'occult chemistry' published in the Theosophical magazine *Lucifer* in 1895. He seems to have replied politely, telling Besant that the book might encourage chemists to search for the missing elements in Dmitri Mendeleyev's periodic table.

At any rate, *Occult Chemistry* did find readers among other scientists, although there is no indication that they took it very seriously. Nonetheless, J. J. Thomson's student Francis Aston, who won the 1922 Nobel Prize in chemistry for his discovery of atomic isotopes (chemically identical atoms with different nuclear masses), borrowed Leadbeater and Besant's prefix 'meta-' to refer to the first known alternative isotopic form of an element, neon-22, in 1912. In 1919 Sinnett published a revised edition of the book that accommodated itself to the new discoveries of nuclear physics, such as radioactivity and isotopes.

The elaborate diagrams of elemental substructures that adorn *Occult Chemistry*, with their lobed shapes classed as 'spikes', 'dumbbells' and arrangements like the Platonic solids, exert a mesmerizing effect and it is not hard to see how contemporary readers might have found this stuff impossible to distinguish from real science. Indeed, the resonances are uncanny: today, the triplets of anu at the atoms' cores seem to echo the three-quark structure of the protons and neutrons that make up atomic nuclei, while the lobes and dumbbells speak immediately to the chemist of the electron clouds called orbitals that were shortly to emerge from quantum chemistry. One can understand why modern mystics are reluctant to accept this as nothing more than eerie coincidence.

Those wicked rays

When X-rays were discovered by Wilhelm Röntgen in 1895 using a modification of the Crookes tube – while 'looking for invisible rays' as he put it – Crookes concluded that there might be etheric vibrations of still higher frequency than these, which could be responsible for the unseen passage of thoughts between minds. X-rays burst upon a *Zeitgeist* peculiarly attuned to such fabulations. Karl Zöllner had just a few years previously quoted Immanuel Kant's speculations about 'other dimensions in space' and 'the existence of immaterial beings in this world', while the philosopher Baron Carl du Prel of Munich, author of *The Philosophy of Mysticism* (1889), wrote in 1892 that '[i]f we imagine other creatures of such infinite aethereal dilution as to be able to penetrate the pores of granite, just as the aether itself can, then the granite would not be there for such creatures'. X-rays seemed to enhance the prospects of discovering such unseen, tenuous realms and their denizens in the ether. There were suggestions in Spiritualist circles that these novel rays might be the odic rays hypothesized by Reichenbach. The author of an interview with Röntgen in *Pearson's Weekly* in April 1896,* in which the German scientist was labelled 'A Wizard of To-Day', claimed that X-rays are 'the last new mystery that human genius has summoned across the border between the known and the unknown'.

Never, said the *Quarterly Review* in April 1896, had a discovery 'so

---

* During the summer of the following year this magazine serialized H. G. Wells's *The Invisible Man*.

completely and irresistibly taken the world by storm'. But it's not clear whether X-rays were afforded a genuinely transcendental role in the public perception, or whether instead these allusions to mysterious realms were offered and received as a mere rhetorical flourish. It has been claimed that Röntgen's discovery spawned an 'X-ray craze' in the late 1890s; but if so, it was a superficial craze for the latest diversion, as transient as any fashion. Berliners could go to the Urania theatre show in 1896 to watch demonstrations and hear lectures on these invisible rays – but its initial popularity waned within a year or so. And while Thomas Edison's X-ray show in New York, which allowed the audience to see the outlines of their bones on a fluorescent screen, left the trick an occult mystery, the Urania performance emphasized that these phenomena had a sound and sober explanation. X-ray tubes became a regular part of the stage magician's equipment; an advertisement in *Nature* in 1896 offered 'Crookes Radiant Matter Tubes' and 'Röntgen Photography' alongside 'Conjuring Apparatus and Mechanical Novelties'.

When X-rays were explicitly allied to the supernatural in public discourse, it was often with a knowing wink. A German newspaper advert for a shop contained a mini-play in which a Frau says to her husband '[h]ave you heard of Röntgen – who, it seems as if a spectre's haunting, has given us rays entirely new?': but the reader is supposed to conclude merely that wonderful things can be bought at the store, not that X-rays are like ghosts. When *Punch* magazine linked X-ray photography with Theosophy in 1896, it was in order to mock both of them: 'go away and photograph Mahatmas, spooks, and Mrs Besant!' There were inevitably outrageous claims for the powers of X-rays – a newspaper in Iowa reported that a young farmer had used them to transmute 'a cheap piece of metal worth about 13 cents to $153 worth of gold', and a French scientist seized on the popularity of X-ray photography to claim to have photographed the human soul. But such reports weren't afforded much credit.

X-rays were described as much with ribald humour as with awe, given their apparent ability to peer through fabrics. One London firm was astonishingly quick off the block, advertising 'X-ray-proof' undergarments for worried Victorian ladies as early as February 1896. *Photography* magazine made light of the 'craze':

The Roentgen Rays, the Roentgen Rays,
What is this craze?
The town's ablaze
With the new phase
Of X-ray's ways.

I'm full of daze
Shock and amaze
For nowadays
I hear they'll gaze
Thro' cloak and gown – and even stays,
Those naughty, naughty Roentgen Rays.

The ability of X-rays to see 'inside' bodies was amusing, titillating and disturbing in equal measure. All of these responses were invoked in George Smith's 'X-ray movie', simply called *The X-Rays* (1897), a comedy in which a courting couple are suddenly reduced to cavorting skeletons (with skeletal umbrella) by the appearance of a crude 'X-ray camera'.

But only after Freud posited the unseen worlds of the mind did anyone truly worry about what these penetrating beams might reveal. 'Nobody knows what other invisible pencils may be registering all our actions or even thoughts – or what's worse, the desires that we don't dare think', suggested the American science popularizer Edwin Slosson

A scene from George Smith's *The X-Rays* (1897).

in 1920. Maxim Gorky speculated about this possibility of using X-rays to photograph thoughts:

> Imagine that someone wants to know you better. He takes [an X-ray] picture of your skull, and if the skull contained some thoughts, the negative will reveal them as black spots, or smokelike spirals, or some other unattractive form. If he wishes, he can try to photograph your conscience, and the negative will also show all the excrescences and blots. In a word, every person will be seen through now, and however thick and impenetrable your skin might be, the new light makes it transparent like glass.

Nuclear phantasmagoria

When Henri Becquerel heard a report of Röntgen's results in January 1896, he wondered whether X-rays might be produced by the phosphorescent substances that Röntgen had used to detect them and which Crookes had used to study his 'radiant matter'. Perhaps these materials could convert light into the new invisible rays? Becquerel was the director of the Museum of Natural History in Paris, where his father Edmond had amassed a large collection of phosphorescent minerals. From among these, Becquerel selected uranium salts, which were known to generate particularly bright phosphorescence. In February Becquerel placed the mineral on top of photographic plates wrapped in black paper and exposed them to light, for as Röntgen had shown, X-rays could penetrate paper to leave an image on the emulsion. Sure enough, that is what Becquerel found when he left the salts and plates in sunlight. But then, having set up another such experiment, he placed the assemblage in a drawer because the February day was gloomy, intending to return to it later. After waiting in vain for several days for better weather, he took out the plates and, through some unaccountable instinct of the sort that often leads to discovery, he developed the plates anyway. There again was the shadow of the uranium salts, even though it had not been stimulated with light. The effect worked in the dark.

So if phosphorescence was producing X-rays, as Becquerel continued to believe, it must be a different, new kind of phosphorescence, indifferent to light. When he heard about the results, Crookes shared that view. It was a curious result but not obviously a profound one, and before 1898 the 'Becquerel rays' were not deemed terribly important.

This changed when Marie Curie, a Polish woman who had come to Paris to study at the Sorbonne and had married the French scientist Pierre Curie in 1895, decided to make these new rays the subject of her doctorate. Discovering that crude uranium ore emitted more of the Becquerel rays than purified uranium salts, the Curies worked to isolate what they presumed was a more 'active' impurity in the ore. After laborious chemical separations lasting four years, they found *two* impurities: both of them new chemical elements, which they named polonium, after Marie's home country, and radium, because its solutions glowed spontaneously with the energy of the Becquerel rays. The Curies renamed these emanations radioactivity. Their efforts, not least their dramatically glowing tinctures of radium, revealed that these invisible rays were in fact far more unexpected and perplexing than X-rays. One didn't need to do anything to these substances to elicit the rays: they just streamed out, and you couldn't stop them.

'Radioactivity evoked a romantic vision', says historian Marjorie Malley. 'It emerged suddenly out of nowhere, invisible and shrouded in mystery.' Marie Curie's accounts of their makeshift laboratory in Paris give a hint of that wonder: 'The glowing tubes looked like faint fairy lights . . . these gleamings, which seemed suspended in darkness, stirred us with ever new emotion and enchantment.'

The unsettling thing about radioactivity was that the energy seemed to come from nowhere. Radioactive substances went right on emitting radiation, regardless of how they were treated chemically, and estimates of the energy they contained gave enormous numbers. The chemist Frederick Soddy declared in 1904 that '[t]he man who put his hand on the lever by which parsimonious nature regulates so jealously the output of this store of energy would posses a weapon by which he could destroy the earth if he chose'. Pierre Curie attested in his Nobel lecture in 1905* how radium 'could become very dangerous in criminal hands'. He went on to warn of the hazards of

---

* For discovering radioactivity the Curies, along with Becquerel, were awarded the Nobel Prize for physics in 1903, but they did not collect the award in Stockholm until 1905 because of illness and the birth of their second daughter.

deep knowledge about nature's hidden mechanisms, using words that could have come from a Renaissance magus of natural magic:

> Here the question can be raised whether mankind benefits from knowing the secrets of Nature, whether it is ready to profit from it or whether this knowledge will not be harmful for it.

To account for the bounty of radioactivity, Marie Curie at first invoked yet more invisible rays. Perhaps, she said, space was filled with radiation even more penetrating than X-rays, which heavy atoms like uranium absorb and then re-emit in another form, much as some minerals absorb ultraviolet light and cast it off as a visible fluorescent glow. That possibility was soon ruled out, however, by experiments showing that radioactivity persisted even in materials taken to the depths of a mineshaft, where the surrounding rock might be expected to screen out even the most penetrating of rays in the environment. The energy source wasn't outside the atom, but inside.

The truth is that radioactivity involves three kinds of 'ray'. As Ernest Rutherford showed in the late 1890s, one of these 'rays' is easily absorbed by aluminium foil: he called this ('for convenience') alpha-radiation. A second, more penetrating ray he named beta-radiation. In 1900 Paul Villars in France showed that there is a third, even more highly penetrating form of radioactivity, called gamma rays, which are electromagnetic waves of even higher frequency than X-rays. Rutherford and others meanwhile established that alpha- and beta-radiation were not rays at all, but particles. The former particle bears a positive charge, and Rutherford showed in 1908 that it is identical to the atomic nuclei of helium atoms. Beta radiation, meanwhile, is nothing other than the negatively charged electron discovered by J. J. Thomson, and presaged by Crookes's radiant matter.

Much of the disruptive impact of radioactivity came from the fact that not only were these 'rays' invisible but so, at first, was their source – for the amounts of radium and polonium that the Curies had to work with were minuscule. 'Chemists were being asked to accept the existence of invisible, imponderable elements on the basis of a phys-ical device which supposedly recorded invisible rays', comments Malley. And yet through this invisible channel streamed energy so limitless that one could be forgiven for suspecting it was a conduit to another, inconceivably potent reality. In 1914 Marie Curie admitted to 'the disconcerting character of the new chemistry . . . of the Invisible which seems to derive from phantasmagoria'– that is, from the spectacular light-show conjurations of the Parisian stage magicians.

Phantasmagoria were certainly the order of the day; with all these unsuspected revelations of immaterial phenomena, nothing seemed too far-fetched to dismiss out of hand. In 1896, shortly after he had reported Röntgen's discovery of the X-ray to the French Academy of Sciences, the mathematician and physicist Henri Poincaré presented to the academicians yet another form of 'invisible light', described in a paper by the psychologist and popular writer Gustave Le Bon, who had found public acclaim the previous year for a book on the irrationality of crowds. Le Bon called this radiation 'black light' and said that, like X-rays, it too affected photographic film placed inside a closed box. For a short time Le Bon's black light created a stir among scientists. But the Lumière brothers, experts in photographic technology, were unable to reproduce his findings, and when Le Bon presented further experiments in 1897 he was criticized by Becquerel as a fantasist. By 1900 most reputable scientists had concluded that Le Bon's black light was an illusion caused by chemical effects in the photographic emulsion. They largely ignored Le Bon's book *The Evolution of Matter* (1905), in which he claimed that all matter was, via the emission of radiation of one sort or another, constantly materializing from and dematerializing into the cosmic ether.

That was by no means the end of fanciful new 'invisible rays'. In 1903 the physicist René Blondlot at the University of Nancy described experiments on X-rays emitted from a cathode-ray tube when a spark passed between the electrodes. When he reduced the voltage applied to the electrodes to the point at which he thought no X-rays were being generated, he claimed to see a change in the brightness of the spark, which he attributed to a new form of radiation that he christened N-rays after his home city. Blondlot said that N-rays would penetrate aluminium and wood, and he later suggested that they were emitted by ordinary gas lamps and incandescent bulbs, as well as by the sun and by steel. Blondlot claimed that the human eye could store and re-emit these rays, while others said that they emanated from the nerves and brains of humans and other animals and even from bodies after death. Inevitably, then, it was only a matter of time before they were being enlisted to explain paranormal phenomena. Le Bon sent Blondlot a letter implying that his earlier studies of black light had identified essentially the same rays, while the German 'psychophysiognomist' and Spiritualist Carl Huter asserted that he was the first to discover N-ray emission from living organisms.

Although other researchers also reported N-ray effects, the problem was that they relied on a subjective assessment of the spark's brightness. The discovery won Blondlot the 1904 Le Comte prize of 50,000 francs from the French Academy of Sciences, but many critics cast doubt on the findings. In late 1904 the American physicist Robert Wood was persuaded to visit Blondlot's laboratory and investigate his experiments. Wood, an expert on the invisible rays of ultraviolet and infrared light, was the ideal man for the job. He quickly established that Blondlot was deceiving himself, at one point spiriting away, without the Frenchman's noticing, a prism that was supposedly crucial for directing the N-ray beams.*

That was the end of N-rays – but by no means the end of the problems posed by invisible rays and particles that could be detected only by their effects.† For physics was now set irrevocably on a path that forced scientists increasingly to infer invisible entities from their visible consequences, whether these were atoms, electrons or photons of light. The problem of occult agencies, which had given Newton so much trouble two hundred years earlier, had come home to roost.

Healing rays

It was suspected at once that X-rays and radioactivity had medical applications. X-rays, for example, could be used to look at bone fractures and to find foreign objects such as bullets lodged in the

---

* Wood later investigated the renowned Italian medium Eusapia Palladino, whose séances were attended by the Curies. He proposed to use a hidden X-ray machine to see if Palladino used her arms to manipulate 'levitated' objects in a darkened cabinet. These tests never happened, however, since Palladino objected to having the arcane equipment of the physics lab intruding on her performance space.

† Invisible emanations continue to be advanced today as explanations for the paranormal. Ultra-low-frequency sound waves, called infrasound and lying just beyond the threshold of normal human hearing, can be produced by some natural phenomena, such as wind and storms, as well as draughts in air-conditioning channels, and have been said to induce feelings of dread and nausea.

body. It was later realized that small doses of radioactive substances can be used to take images of organs in which they become concentrated, or as tracers to track ingested material as it circulates in the body: you simply use a detector to find out where the radioactivity is being emitted. Soon enough the dangers also became apparent. Thomas Edison's assistant in X-ray work, Clarence Dally, developed cancers in his hands that required both arms to be amputated, although the unfortunate man died soon after in any case. From their studies of radioactive materials the Curies suffered radiation burns, and both Marie and her daughter Irène died of radiation-induced blood conditions. Irène's husband Frédéric Joliot, another pioneer in the emerging field of nuclear science, also died prematurely because of radiation exposure. Despite the carcinogenic potential of X-rays and radioactivity, however, both can be directed towards the selective destruction of tumours, for which purpose they are still used today.

At first the health effects of the new rays were thought to be purely benign. After all, occult forces and emanations have always been attributed miraculous healing powers, from the medieval use of magnets to treat anything from gout to baldness to Mesmer's animal magnetism; there was no reason why X-rays and radioactivity would be any exception. Dubious X-ray therapies began to appear in the early twentieth century, while radium cures were offered as panaceas for every ailment; 'therapeutic' radium was even added to toothpastes and cosmetics. The publicity for these radioactive remedies had the fervour of faith-healing: 'Whatever your Ill, write us', said the Nowata Radium Sanitarium Company in 1905, 'Testimonials of Cases cured will be sent you.' The company's advertisements showed Native American maidens beckoning alluringly from a picture of the globe 'encircled with radium water'. For all the trappings of science, an association with ancient, 'primitive' healing lore was evidently still useful.

Spas boasted of it proudly when radioactivity, from natural radioactive elements such as radon in rocks, was detected in their waters. Bath in England, Baden-Baden in Germany and Hot Springs in Arkansas all advertised their radioactive elixir. St Joachimsthal, near the largest uranium mines in Europe, became a spa town, while in Claremore, Oklahoma, a cluster of hot-spring bath-houses became known as Radium Town.* Even today, Bad Gastein in the Austrian Alps publicizes its radon-laced water, and indeed some researchers propose that very low doses of radioactivity might have health benefits by stimulating the body's protective mechanisms in a manner somewhat akin to vaccination.

That the magical legacy of 'invisible healing agencies' was in the air in the early days of radioactivity is clear from a newspaper report of 1909 describing the discovery of a 'new' radioactive substance (a 'rival to radium') by Dr Skillman Bailey of the Hahnemann Medical College in Chicago,† which he called 'radio-thor':

> It possessed, said Dr Bailey, all the curative properties of radium, and none of its baneful after-effects. It is also within the reach of persons of moderate means, and the supply is apparently limitless . . . Dr Blackmarr, of Chicago, who has been associated with Dr Bailey in laboratory experiments, declared that the discovery is of such great importance to humanity and the medical world that he hesitated to express himself adequately lest he should seem to be going beyond the mark . . . 'It is idle to attempt an enumeration of the ailments that radio-active applications will cure,' said Dr Blackmarr. 'In fact, what we really wonder is whether anything exists that it will not cure once we have thoroughly mastered the handling of it. I should not like to suggest that we have at last found a means for the indefinite prolongation of human life by arresting the processes of decay, yet it is a fact that the experiments we have made indicate extraordinary power in the new agent to prolong life.'

Radio-thor was indeed radioactive and Bailey's potion was lethal to some patients – including one Eban Byers, whose death from radiation poisoning helped to trigger regulation of radium therapies in the

---

* Radium was not actually detected in the Radium Town waters until the 1950s, by which time its dangers were appreciated. The original claim was based only on the sulphurous nature of the waters. This was not uncommon for spas advertising 'radium water'.

† It's no surprise to learn that the 'Hahnemann' of the college's name refers to Samuel Hahnemann, the eighteenth-century inventor of homeopathic medicine, and that the college actively promoted his ideas. If an (infinitely small) amount of a toxin does you good, might not that work for radioactivity too?

1920s. But the restrictions came too late for the many factory workers of the US Radium Corporation killed or incapacitated by the radium paint that they applied to the luminous dials of clocks and watches in the first two decades of the century.

New worlds

The invisible rays of the late nineteenth century created a new physics in which there was ample room for the para-physics of telepathy and telekinesis, the play of yet more unforeseen forces and the possibility of concealed planes of existence with their own invisible denizens. The ether, the tenuous medium of the vacuum, seemed suddenly to be surging with activity: it could certainly carry voices across the Atlantic, and (who knew?) perhaps across the Styx. There was more energy hidden away inside the atomic nucleus than filled the spaces outside it. Some scientists suspected that our visible world was a phantom, an illusion ('albeit a very persistent one', as Einstein is said to have put it) created by the sensations that invisible force-fields activated. What looked like solid particles might be simply knots in the ether. The dazzling dance of light inside Geissler discharge tubes prompted scientists to wonder if matter might be nothing more than a ghostly manifestation of this evanescent energy. 'Around the turn of the century, the option that electromagnetism was more fundamental than matter looked promising', says Malley.

Everywhere there were signs of invisible activity, the operation of forces and processes we could only glimpse indirectly – via shadows on photographic plates, glowing glassware, noises, voices, spinning windmills and dark voids carved in light. In the Renaissance, exploring new worlds had demanded a voyage into the unknown. Now that voyage was headed into the unseen.

5 Worlds Without End

It is not a very incredible thing to suppose that in the luminiferous ether (or in some other unseen material medium) life of some kind exists.

William Barrett
On the Threshold of the Unseen (1917)

It is not difficult to design thought-experiments which demonstrate that scientists could easily form an inadequate idea of the extent of physical reality.

Frank Wilczek, Nobel laureate in Physics, 2013

William Barrett was puzzled by flames. Barely out of his teens, he was assisting John Tyndall at the Royal Institution in London in the 1860s when he noticed that flames seemed to be sensitive to high-pitched sounds. They would become flattened and crescent-shaped, as Barrett put it, like a 'sensitive, nervous person uneasily starting and twitching at every little noise'. He was convinced that this 'unseen connection' – 'so very magical' – was not mediated by anything material and tangible. It was, he admitted, an effect 'more appropriate for a conjuror's stage than a scientific lecture table.'

Barrett came to regard certain people as analogues of the 'sensitive flame', exquisitely attuned to immaterial vibrations that others could not perceive, to 'forces unrecognized by our senses.' These persons could read the thoughts of others (Barrett had witnessed mind-reading demonstrations by a young girl), and they were able to receive messages from spirit-beings that, in Barrett's view, were not exactly supernatural but merely 'supernormal', existing in an intermediate state within the ether between the physical and the spiritual.

It is no surprise, then, that Barrett was attracted by William Crookes's psychical researches in the 1870s. Although initially Barrett suspected mediumistic phenomena to be the result of hallucination, his own experiments on telepathic communication of letters, words, names and playing-card faces between hypnotized subjects convinced him that the claims of Spiritualism warranted serious scientific study. In 1881 Barrett, now Professor of Physics at the Royal College of Science in Dublin, published his findings on thought-transference in *Nature*. The five children of a Derybyshire clergyman, he reported, were able to transmit such information mentally between one another, with a failure rate of not more than one in ten trials. The ensuing controversy about these experiments motivated Barrett to convene a group of like-minded individuals who would conduct psychical research as a systematic science.

The Society for Psychical Research (SPR), a diverse collection of academics and Spiritualists, arose from a meeting between Barrett and Edmund Dawson Rogers, vice president of the Central Association of Spiritualists, in 1882. Crookes duly enrolled; he joined the Council in 1891 and remained on it until his death in 1919, becoming the society's president in 1897. Here at last was a forum where he could safely profess his convictions about the reality of invisible beings of 'intelligence, thought and will, existing without form or matter, and untrammelled by gravitation or space.'

The SPR is an odd beast (it still exists). On the one hand it turns a sceptical light on paranormal phenomena, accepting that they might be no more than psychological, and it offers a forum for scholarly historical studies of the field. On the other hand it is receptive to reports and theories that are strange, vague, speculative and most definitely on the scientific fringe. That has always been its nature. Its first president, Henry Sidgwick, was Professor of Moral Philosophy at Cambridge and doubtful about the claims of Spiritualism. Other presidents have included William James, Lord Rayleigh and Arthur Balfour (later prime minister of Britain), and among its members were J. J. Thomson, Lewis Carroll, Alfred Tennyson, John Ruskin and the former prime minister William Gladstone. Today, membership is apt to place a question mark against one's reputation as a scientist, although the scientific credentials of astrophysicist Bernard Carr, president for 2000–2004, are impeccable. The problem is the same as that faced by Faraday when he studied table-turning: it is considered one thing to be a sceptic of the paranormal, and quite another to be prepared to put your scepticism to the test.

Like Crookes, Barrett suspected that at least some psychical phenomena might be explained as the interventions of invisible, immaterial beings – not souls or ghosts, but natural, living creatures. In his apologia for psychical research *On the Threshold of the Unseen* (1917), Barrett described them as 'human-like, but not really human, intelligences – good or bad *daimonia* they may be, *elementals* as some have called them'. Allow this much, and anything can follow. Why shouldn't these *daimonia* be responsible for 'the passage of matter through matter, or the knotting of a single endless cord, or loop, or ring of leather' – feats that mediums could allegedly accomplish? Yet if this was modern demonology, it hadn't entirely lost a certain medieval aspect: 'mischievous agencies doubtless exist in the unseen', Barrett warned, and it was necessary to be on guard against their 'invasion of our will'. These spirits were – for better or worse – invaders from another realm. But from where, exactly?

The scientific soul

One potential answer was proposed by the Irish physicist Edmund Edward Fournier d'Albe, who like Barrett taught in Dublin until moving in 1910 to the University of Birmingham in England. Fournier d'Albe was interested in electromagnetic phenomena and conducted experiments in radio and nascent television technology; he was a regular writer in the *Electrician*. Like Crookes* (albeit with a more doubting view of Spiritualism) he merged these interests with a belief in invisible beings and worlds with which we stood on the verge of making contact.

In *Two New Worlds* (1907) Fournier d'Albe argued that the recent discoveries in radioactivity and atomic structure implied the existence of an unseen spiritual universe contiguous with ours. As the physicist George Johnstone Stoney (who coined the name for J. J. Thomson's

---

* Fournier d'Albe published a biography of Crookes in 1923.

electron) had recently argued, the material universe must now properly be regarded as an infinite series of worlds within worlds, which Fournier d'Albe considered to differ only in the size of their elementary constituent particles. He discussed two of them, the 'infra-world' of atoms and electrons (see page 211) and the 'supra-world' of cosmic proportions: descendants of the Neoplatonists' microcosm and macrocosm. And as above, so below: both are, like our own world, teeming with purpose and life.

Fournier d'Albe expanded on these views the following year in *New Light on Immortality*, where he tried to come to terms with what the notion of a human soul could possibly mean in the atomic age. To pronounce on immortality, he said, who now was better placed than the physicist, who understood the most about energy and matter? Suppose that what we call the soul is a substance more tenuous than vapour, composed of particles integrated into our body but in principle separable from it. Fournier d'Albe called these structures 'psychomeres', and suggested that they possessed a kind of intelligence and ability to interact with each other via telepathic contact.

Although in the present state of science 'we cannot reasonably expect to see psychomeres', Fournier d'Albe admitted, he felt able to deduce something of their nature. Actually it was sheer guesswork – estimating the number of psychomeres comprising a human soul, for example, he plucked a figure of ten trillion out of the air. From this he calculated the mass of a soul to be about 50 milligrams,* and asserted that, were the soul-matter of a person to be condensed into a body just six inches high, it would have the same density as air and would float freely in it. Such a concentration of psychomeres might border on visibility: it could resemble a will o'the wisp or one of the tiny sprites ubiquitous in folkloke. 'And thus it comes about that all the fairies, pixies, sylphs, and gnomes fly before the flaring light of science', Fournier d'Albe proclaimed triumphantly. 'They are not so much sent away as explained away.'

---

* This is appreciably less than the popular estimate of 21 grams, which derives from the work of an American physician, Duncan MacDougall, published a year before Fournier d'Albe's book. MacDougall arrived at this figure after weighing human bodies just before and just after death. The numbers testify to an increasingly materialistic view of souls among some of those who still believed in them.

We might expect a liberated agglomeration of psychomeres to adopt a streamlined shape for ease of progress through the air – like a fish, which Fournier d'Albe points out was a symbol of early Christianity, or like a flame, as seen on Moses' burning bush. And suppose if, once a soul has left the mortal body, its 'earth memories . . . should be awakened, and become dominant' – then it might gather again into its remembered earth-form:

> First, a fine mist, then a cloud, a tall pillar of filmy vapour, from which a complete form, moulded and clothed to suit the character assumed,* would then emerge, to walk the earth as before for a little while, and to dissolve again into mist and become once more invisible. The inhabitants of the earth would then see a ghost, and be afraid.

According to this theory, there was no longer any reason to fear these manifestations, for they were us, disembodied: spirits with an intelligence 'no higher than the average human being, neither devil nor angel'.

Today's observer might consider these ideas to be unmotivated speculation, if not little more than a tautological redefinition of the traditional soul. There is, after all, not the slightest reason to suppose that any fundamental particles have intelligent autonomy and telepathic communication. But Fournier d'Albe might have argued that he was assaying no more than what science had always done: to reduce complex, puzzling phenomena to a minimal set of propositions, based on mechanical principles, that could rationalize them. Besides, invoking an invisible world might offer consolation for an increasingly barren picture of the physical realm. From natural history, he wrote,

> theology has been ruthlessly evicted. The visible world being henceforth closed to it, it has taken refuge in the invisible world, where it

---

* Fournier d'Albe addresses the common jibe against Spiritualism of why spirits should appear clothed, when their garments are mere earthly apparel and irrelevant in the afterlife. If psychomeres can reconstitute the human form, he says, 'why draw the line at a little additional "ballast" which enables the forms to appear in a mixed company without immediately raising insuperable objections to their presence?' It was considerate of the laws of the invisible universe thus to respect Edwardian decorum.

feels free to make what declarations it likes. And that invisible world continues to be the 'home' towards which the weary heart turns from a world that has become indeed clean and bright and sanitary, but utterly hopeless and empty, if not unjust and cruel.

And yet, he says, this flight to an invisible world should not remove us from the world we tangibly occupy:

> We must resolutely combat the tendency to look for the unseen beyond the seen. The unseen is all about us . . . a single octave in the gamut of light-waves impresses our retina, revealing a very small proportion of what would be visible to a more completely equipped intelligence.

X-rays had made that plain enough.

Theological thermodynamics

We can see, then, that the idea that there might be an entire realm of existence, immaterial yet peopled with intelligent agents who could contact and interact with our physical world, was emboldened by the new discoveries of the *fin de siècle*. As Marjorie Malley puts it,

> The invisible force fields of electromagnetism, mysterious electrical discharges, the invisible ether, and discoveries of invisible [ultraviolet] light and other radiations reinforced speculations about an unseen ghost world that could be contacted by sensitive intermediaries, or mediums. These were not fringe ideas.

They certainly were not; Barrett, like Crookes, possessed a knighthood and a Fellowship of the Royal Society. And although these speculations might seem now to be an extraordinarily expensive way, in terms of the novel and untested principles they demanded, to explain the odd goings-on in the houses of mediums, we should remember that most of that conceptual cost was already paid by Christian faith. If some nineteenth-century scientists, such as John Tyndall and Thomas Henry Huxley, started to question the price, the majority considered it inevitable. As a scientific understanding of the world advanced, some scientists still felt a need to reserve a space for God, the soul and the afterlife. No telescope or microscope was going to locate these things; they would have to be invisible.

The most notable and fully developed effort to provide a scientifically plausible account of invisible spirit worlds within a Christian context was made by the distinguished Scottish physicists Balfour Stewart and Peter Guthrie Tait in their book *The Unseen Universe* (1875). Although Stewart became president of the SPR during the 1880s, both men were sceptics of Spiritualism, seeing in it nothing more than evidence of human suggestibility, 'the power that one mind has in influencing another'. Tait attacked Spiritualists at the British Association meeting of 1871, bracketing them alongside 'Circle-squarers, Perpetual-motionists [and] Believers that the earth is flat'. Yet he and Stewart were eager to understand how the 'invisible order of things' that the Bible seemed to require – the existence of immortal souls – might be consistent with the laws of physics. They insisted that 'the presumed incompatibility of Science and Religion* does not exist'. And yet their conception of religion, at least on the evidence of *The Unseen Universe*, was starkly materialistic: they fit within a long tradition of both advocates and opponents of religion who insist on making it a set of beliefs about the physical world that may either be rationalized or disproved.

In truth, however, the questions that Stewart and Tait tackled are profound ones that continue to vex cosmologists today. If the universe has a beginning in time and an end in space, what lies before and beyond? And why has this universe produced something as seemingly unlikely as ourselves – creatures able and indeed compelled to ponder their origin? 'The visible universe must, *certainly in transformable energy, and probably in matter*, come to an end', they wrote. 'We cannot escape from this conclusion.' But discontinuities in time and space do not make logical sense, and so

> we are forced to believe that there is something beyond that which is visible . . . an invisible order of things, which will remain and possess energy when the present system has passed away.

* This is a reference to John Tyndall's notorious attack on religion in his address to the British Association in Belfast in 1874, in which he asserted that religion should not be permitted to 'intrude on the region of *knowledge*, over which it holds no command'.

This unseen realm need not be remote, but may be present right alongside us, within reach if only there was anything to touch, and populous too. Its fabric might lie at the extreme of the gradual dematerialization of substance we see already in the physical world, where solid, liquid and vapour give way to the 'semi-material' existences of electricity, magnetism, heat (hypothesized by some scientists to be a substance, called caloric), light, and the occult field of gravity.

The invisible universe, Stewart and Tait argued, must have existed before the visible one, and must be capable of acting upon it – indeed, of producing it. Life itself is a 'peculiarity of structure which is handed over . . . from the invisible to the visible . . . In fine, life as well as matter comes to us from the Unseen Universe'.

This bounty offered by the invisible to the visible world relied on communion between them. And what should provide that contact but the rainbow bridge of nineteenth-century physics, the ether? 'May we not regard ether', they wrote,

> as not merely a bridge between one portion of the visible universe and another, but also as a bridge between one order of things and another, forming as it were a species of cement, in virtue of which the various orders of the universe are welded together and made into one? In fine, what we generally call ether may be not a mere medium, but a medium *plus* the invisible order of things, so that when the motions of the visible universe are transferred into ether, part of them are conveyed as by a bridge into the invisible universe and are there made use of or stored up.

This ether-mediated communication is vital to the authors' theory of the immortality of the human soul. For we each possess a 'spiritual body' in this invisible world, which becomes energized by our actions and impulses in the tangible world. 'Certain molecular motions and displacement in the brain', Stewart and Tait wrote, are in part 'communicated to the spiritual or invisible body, and are there stored up' as a kind of latent memory. This accumulated energy makes the spiritual body 'free to exercise its functions' even after bodily death. By living, we store up our immortality. For all we know, the universe might be teeming with thousands of these spiritual worlds, 'some existing in different parts of space, others pervading each other unseen and unknown, in the same space, and others again to which space may not be a necessary mode of existence'.

There is, however, a problem. In 1850 the German physicist Rudolph Clausius formulated the first and second laws of thermodynamics. The first law says that energy is never destroyed but always conserved, being merely converted from one form to another. The second law says that heat always flows from hot bodies to cold ones, never the reverse.* A year later William Thomson pointed out that the flow of heat involves 'dissipation' of mechanical energy: it flows into random motions of molecules, at least some of which can never be recovered to do anything useful. This process, he said, must eventually create a universe of uniform temperature, from which no useful work can be extracted, and in which nothing really happens. But how can this 'heat death' of the universe be consistent with immortal souls?

Here Stewart and Tait fell back on an idea proposed by their mutual friend James Clerk Maxwell, who worried about the implications of the second law of thermodynamics for human free will. How can God's gift of human freedom to act (and thus to choose our salvation) be reconciled with a universe heading inexorably towards an inert, lifeless state? Maxwell's solution was first articulated in a letter to Tait in 1867. What if, he said, there exist invisibly small beings – later dubbed 'demons' by Thomson – that could cheat the second law by identifying 'hot' atoms and separating them from 'cold' in a random mixture, creating a reservoir of heat that could be tapped to do work (see pages 207–8). Such beings, Stewart and Tait now proclaimed, might 'restore energy in the present universe without spending work'. It isn't clear whether Maxwell ever intended his demons to be more than hypothetical. But for Stewart and Tait they became essential agents of eternal life.

Evidently, for these authors the unseen universe was an agent of all possibilities. 'The scientific difficulty with regard to miracles will, we think, entirely disappear, if our view of the invisible universe be accepted', they claimed. Indeed,

---

* More precisely, there can be no *overall* flow of heat from cold to hot. It is possible to arrange such a local flow – that's how fridges work – so long as the flow is compensated for by hot-to-cold transfer elsewhere (which is why the fridge needs to be plugged in, and is warm at the back).

It appears to us as almost self-evident that Christ, if He came to us from the invisible world, could hardly (with reverence be it spoken) have done so without some peculiar sort of communication being established between the two worlds.

This, then, is the direction in which invisible forces and rays pointed for some scientists in the late nineteenth century: towards what we might regard as a thermodynamic theory of God and Christ, as well as (for evil might not be limited to the visible world) of eternal Hell. *The Unseen Universe* culminates in a bizarre expression of piety, couched in the language of Victorian physics:

[Man] ought to live for the unseen . . . But, in order to enable him to do this, the unseen must also work upon him, and its influences must pervade his spiritual nature. Thus a life *for* the unseen [God?] *through* the unseen is to be regarded as the only perfect life.

Perhaps concerned about how far they had pushed their ideas, Stewart and Tait published their book anonymously.

The hidden reality

We have never looked back from this dematerialization of the world that began a century and a half ago. The speculations of Crookes, Barrett, Fournier d'Albe, Stewart and Tait were proposals that our visible world is not the only reality. This is just what physicists today still assert with their notions of strings and extra dimensions. Contemporary metaphors such as a 'hidden reality' recommend them-selves precisely because they have a history. Can there be any doubt that Spiritualists would have delighted in 'dark matter' and 'dark energy', these unseen particles and forces that supposedly dwarf the meagre quantities of visible matter in the universe and propel it on a trajectory that gravity is powerless to counteract? When, in describing such concepts, cosmologists speak of 'unravelling the mysteries of the invisible universe', as two of their leading representatives have done recently, they are invoking a legacy that, if they knew of it, they might not welcome – but which they cannot deny.

None of this is to cast aspersions on the motivations behind these ideas. Dark matter and dark energy – both of them invisible agencies with cosmic roles to play – are inferred, in fact made necessary, by their apparent effects on the visible universe: a secular version of the Biblical contention (Romans 1:20) that 'God's invisible qualities . . . have been clearly seen, being understood from what has been made' (that is, from what can be seen). Astronomers are sure that dark matter must be present in the universe, because without its gravitational effects rotating galaxies should fly apart – the stars and gas towards the edge of spiral galaxies rotate faster than should be possible if one considers only the masses of the visible matter.* It is 'dark' because it doesn't seem to interact directly with light or any other electromagnetic radiation, which implies that it is unlike any form of matter already known. Several new types of particle have been proposed to account for it, all of them speculative and motivated by nothing more than necessity. But despite various experiments to detect them, and hopes that they might even be created in the mightiest particle accelerators, we still know next to nothing about what dark matter is.

Dark energy is another name for a concept that is mysterious, invisible and, again, invoked only because there seems to be no alternative. It is believed to permeate all space and to act as a kind of negative gravity that is causing the expansion of the universe to accelerate. Until the 'discovery' of dark energy in 1998 it was widely assumed that, once the Big Bang had banged, this expansion would gradually slow because of the braking effect of gravitational attraction between all the matter in the universe. But observations of distant exploding stars called supernovae show that in fact the expansion is speeding up.

This was a 'discovery', however, only in the sense that it made the existence of something like dark energy impossible to ignore. Cosmologists had known for years that space *should* be intrinsically 'repulsive'. Attempts to express quantum mechanics as a theory involving underlying, invisible fields (called quantum field theory) suggest that empty space should in fact be so full of energy (called vacuum energy) that the existence of ordinary matter would be

---

* There are other, independent reasons for invoking dark matter in galaxies too. One of the most compelling is that it appears to bend light, just as Einstein's theory of general relativity predicts for any substance, so that the images of still more distant galaxies get distorted.

impossible. Yet we don't see all this vacuum energy or its effects (if it was there, we wouldn't be), leading most researchers to assume that for some reason it is cancelled out exactly. The discovery in 1998 that the expansion of the universe is speeding up suggested that this cancellation of the vacuum energy is not quite complete: a little bit remains to drive the accelerating cosmic expansion. The odd thing about dark energy, then, is not that it exists, but that there is so little of it. The tiny amount that remains is so close to zero as to defy all intuition: if some factor cancels out the vast quantity of vacuum energy that quantum field theory predicts, why is that cancellation so very, very nearly – but not quite – exact?

'Tiny' is relative here, for dark energy is thought to account for about 68 per cent of all the matter/energy* in the universe. Dark matter accounts for a further 27 per cent, leaving just 5 per cent or so that is made up of all the matter and energy we can actually see. In this sense, it seems that there certainly is an invisible universe and that our visible universe is rather insignificant in comparison.

Not all of today's 'hidden' universes are invisible in the usual sense. Some are thought to be merely out of reach – but fundamentally so, too distant for us to journey there even if we could travel at the speed of light. If the universe is infinite, and if the part of it we can see is representative of what lies beyond, then we can regard it as a patchwork of mutually inaccessible regions in which, logically speaking, every conceivable permutation of particles must be realized sooner or later: endless copies, differing from our own 'universe' in every trivial and non-trivial way. Or perhaps our 'local' universe is not representative after all, but merely one example among countless variations of how the laws of physics might turn out. This vision seems permitted, perhaps even compelled, by the current favourite theoretical scenario for the Big Bang, which invokes a period of extremely rapid expansion called inflation. The idea here is that small differences in the fundamental constants of physics from one place to another in the nascent, microscopically small universe – random fluctuations, like variations in air pressure throughout the earth's atmosphere – get suddenly blown up by inflation into entirely distinct universes where different rules apply.

---

* Einstein's theory of special relativity makes matter and energy equivalent.

The attraction of this notion of 'bubble universes' is that it could account for why the conditions in our universe – the fundamental constants of physics – seem fine-tuned to precisely the values needed for stars, galaxies and, ultimately, life to appear.* We needn't invoke some supernatural foresight or planning to account for our presence; rather, we are able to ponder this 'fine-tuning' at all only because we inhabit one of the few universes, among innumerable alternatives, in which this situation obtains. Some scientists reject this so-called 'anthropic' reasoning entirely, asserting that it is close to being a circular argument and not obviously testable against any observations. But alternative suggestions for how fine-tuning might happen – one hypothesis posits a system of universes replicating via the formation of black holes and undergoing a kind of natural selection that favours universes abundant in stars (and thus in black holes arising from their collapse) – also typically assume that there exist multiple universes, or a *multiverse*.

Yet all these 'hidden worlds' are invisible only in the sense that China is invisible from my back garden: the issue is geographical. More unnerving, perhaps – but also more familiar within the history of unseen universes – is the speculation that we *coexist* with invisible worlds, less than a hair's breadth away, superimposed on our reality and accessible if only we had a power, denied to quotidian existence, to reach through new dimensions or touch new spaces. These are not the fantasies of cranks. Mindful of the inhabitants of Edwin Abbott's classic popular science book *Flatland: A Romance of Many Dimensions* (1884),† who are confined like ants to a two-dimensional plane and can't conceive of, let alone reach, other flat universes hovering just above or below, the Nobel laureate physicist Frank Wilczek has asked '[a]re we human beings comparably blinkered?' After all, he warns, 'in the past scientists have repeatedly reached "intellectual closure" on inadequate pictures of the universe, and underestimated its scale.'

---

* One aspect of this fine-tuning involves the invisible entity of dark energy itself, specifically why there is so little of it.

† Abbott, a schoolmaster, intended his short novel to be a social satire. It was largely ignored until physicists 'rediscovered' it after Einstein's theory of general relativity in 1915 had introduced the idea of four-dimensional spacetime. This rediscovery owed much to a reference to the book in a letter to *Nature* in 1920 from William Garnett, who worked as a demonstrator at the Cavendish Laboratory in Cambridge under James Clerk Maxwell and subsequently published a biography of Maxwell.

It is certainly possible to *imagine* alternative universes separated from ours by extra dimensions beyond the three spatial dimensions in which we are confined. (Or rather, it is possible to postulate them; whether we can picture quite what this means is another matter.) But if these universes are then by definition beyond our ability to see or to detect from any observable effects in our own universe, why should we wish to suppose that complication? Physicists have their reasons. The first intimation of hidden dimensions began with the work of the theoretical physicist Theodor Kaluza in the early twentieth century, who showed that by adding an extra dimension to the equations of Einstein's theory of general relativity he could obtain an additional equation that seemed identical to Maxwell's theory of electromagnetism and light. Since general relativity provided a framework for understanding gravity, Kaluza's result seemed to suggest a link between gravity and electromagnetism – a step towards the grand unified theory of fundamental forces that Einstein was then seeking. After initially ignoring the work when Kaluza sent it to him in 1919, Einstein finally realized its potential significance and helped Kaluza to publish it.

But where, then, was this extra dimension? The Swedish physicist Oskar Klein offered an answer in 1926: perhaps it was curled up into an absurdly small length of around 10^{-33} cm, known as the Planck length. The clichéd but indispensable comparison here is with a garden hose: a three-dimensional object which looks like a one-dimensional line from far away because the other two dimensions are so small. It takes no significant time at all to cross Klein's extra dimension, so we don't notice it.

An extra dimension as wide as the Planck length hardly seems a place into which we can squeeze another universe. But a hypothesis like that of Kaluza and Klein becomes much more extravagant in string theory, which seeks to explain fundamental particles as the vibrations of even smaller entities called strings. When string theory was developed in the 1980s, it was found not just to permit but actually to require extra dimensions beyond the familiar four of spacetime – and not just one extra, but up to seven in the extension of string theory known as M-theory. (It has always been left ambiguous what 'M' stands for – some say 'magic'.) What is more, these dimensions need not be compact at all, but can be extended objects called branes (a contraction of 'membranes'), which may be multi-dimensional. A brane might be a perfectly adequate hiding place for a universe, and indeed M-theory postulates a multiverse of branes of various sizes, coexisting rather like a stack of papers. In general these brane worlds remain quite distinct and separate from each other, because the four fundamental forces of nature do not reach between them,* and so each is unobservable to the other. But if branes do touch and collide, the result could be catastrophic: our own Big Bang might have been triggered that way.

Here, then, are invisible universes aplenty, more than you can imagine, separated from our own perhaps by a distance that barely registers – but which extends in a dimension forbidden to us. As physicist Brian Greene puts it, there may be 'giant branes right next door, entire parallel universes hovering nearby like slices of rye cozying up to their neighbors'. But to reach between worlds one would need a subtler knife than that which opens portals in the membranes between worlds of Philip Pullman's *His Dark Materials*.

The problem is that there is not the slightest evidence that this is actually how things are. Some physicists believe that the mathematics of M-theory gives us no alternative but to accept the idea of branes. Others deny that such predictions amount to science at all, for they cannot be tested by any known experiment. They are a leap of mathematical faith, predicated on so many hypotheticals and conditionals that they have been dismissed as a form of metaphysics, even quasi-theology.

And that's not the end of it. Quantum theory, which shattered Maxwell's smooth electromagnetic waves into fragments called photons, spawns another welter of coexisting but invisible worlds, with what one can only call indecent fecundity. Here's why. In 1926 the Austrian physicist Erwin Schrödinger wrote down an equation describing how the behaviour of a quantum system – an electron in an atom, say – evolves over time. Schrödinger's equation might be seen as the quantum equivalent of Isaac Newton's equations of motion: if you know the situation that the particle is in (the forces that act upon it) at some instant, you can calculate its trajectory at any time in the future. But what the Schrödinger equation gives you

---

* Actually one proposal is that gravity is unique among these forces in being able to 'leak' between branes, explaining why it is so weak in our universe relative to the other forces.

is not exactly a trajectory, which would be an indication of where the particle will be and how fast it is moving at a given point in time. Instead, it tells you the probability that the particle will have particular values of these things. In the quantum world that's all you can know: not how things will be, but how likely they are to be this way or that. The Schrödinger equation is a prescription for the various options.

This is perfectly fine, except that if you actually make a measurement on the particle then you find just one result, not a range of possibilities. The electron is either here or there. What happened to all the other possibilities that the equation permits? According to the so-called Copenhagen interpretation of quantum mechanics, developed in the 1920s by Niels Bohr and his collaborators in Copenhagen, measurement causes all the solutions of the Schrödinger equation to collapse into just one outcome – the others are in effect thrown away, at least for the purposes of that particular measurement. But there is nothing in the mathematics of quantum theory that anticipates or describes this collapse: you just have to put it in by hand. That doesn't seem very satisfactory. This is the so-called 'measurement problem' of quantum mechanics.

One resolution to this problem was proposed in 1957 by the American physicist Hugh Everett. Perhaps, he said, all those other possible states allowed by the Schrödinger equation don't go away. His radical suggestion was that there was no collapse of the wavefunction at all: each of the alternative solutions continues to exist as a separate reality – even though we can observe only one of them.

This has become known as the Many Worlds interpretation of quantum theory. You can doubtless see why that is, but Everett himself never called it by that name. He didn't really address the question of where those extra states reside. He certainly did not take the extraordinarily profligate, and theoretically unmotivated, step of constructing an *entire parallel universe* around them, identical in all respects except for the outcome of the single quantum event in question. His idea was a mathematical formalism; the 'Many Worlds' proposal, in contrast, is an interpretation of what it means.

The popular image of this view of quantum theory is of a single reality splitting into two (or more) when the measurement is made. (And let's be clear that 'measurement' doesn't mean a scientist observing something in the lab, but any interaction that affects the particle, such as a collision between two atoms, or between the particle and a quantum fragment of light – in other words, things that happen all the time, everywhere, in more ways within a single second than you could ever hope to enumerate.) But the notion that this introduces new branches of reality is not quite right. In fact those alternative realities were already inherent in the initial state of the system, which is to say, in the initial Schrödinger equation. In the Many Worlds interpretation of quantum theory there is really only one über-Schrödinger equation describing the entire quantum multiverse, and its corresponding solution or 'wavefunction' flows through time like a great braided river, all its possible realities evolving at once in an incredibly complex interplay. All coexist in the same space, and all are made of the same materials, but once branching has happened, they can't communicate with each other.*

There are several leading physicists – Wilczek is one of them – who insist that there is no option but to accept the Many Worlds interpretation as the only vision of reality that takes the maths of quantum theory seriously, and does not impose arbitrary contingencies such as collapse of the wavefunction. This, like the invocation of a brane multiverse, is a statement of untestable faith, and a weirdly circular one at that. Wilczek proposes that the real-world 'evidence' for the Many Worlds hypothesis is the very issue that motivates it: the fact that it does not demand wavefunction collapse. The truth is that how we go from Schrödinger's equation of multiple probabilities to the observation of an apparently unique reality is a mystery, for the solution of which there are several options (and probably others not yet identified) – for example, some scientists imagine that this collapse might be a real, physical phenomenon, rather like radioactive decay. The idea that an abstract mathematical entity like the Schrödinger equation must be our ultimate descriptor of reality, when it lacks any truly fundamental justification itself (although it works extremely well in practice), represents a capitulation to formalism that seemingly only physicists have the capacity to indulge. To say 'this is the best theory we can currently come up with, so it must be true'†

---

* This issue of whether, after 'splitting', any of the 'many worlds' can interact with each other in the future is in fact a little hazy, although this can only happen if any memory of the initial splitting is lost – so that there's no chance of detecting an alternative reality.
† Actually even this is not right, for some researchers are developing a form of quantum theory that does not require the Schrödinger equation, although it permits it as one way of doing the maths.

is the kind of blinkered assumption that Wilczek rightly points out has misled scientists in earlier ages.

But it's worse than that. For we must ask: where are *we* in the Many Worlds? And the only answer one can give is that we are everywhere. The only meaningful interpretation of *we* is an entity present in every one of the many worlds at once, experiencing all things that it is possible to experience, all possible solutions of the equation. There is no perpetual splitting of ourself into distinct entities that exist in the separate worlds: everything can and does happen to us that could conceivably happen. 'The act of making a decision causes a person to split into multiple copies,' asserts physicist Max Tegmark – but Greene adds that 'each copy *is* you', evincing scant concern for the metaphysical (not to mention legal) conundrums this poses.*

The problem with statements like this is that they are deceptively easy to write – but do you see what they mean? You do not – because they don't have a meaning that can be articulated. This is not a coherent definition of selfhood, but a clumsy stitching together of the notion of a wavefunction and the notion of mind that makes the crude requirement of wavefunction collapse look trivial in comparison. As a solution of the measurement problem, the Many Worlds hypothesis has clear attractions. But once one moves beyond the kind of refined situation that quantum theorists tend to think about, with individual particles passing along this pathway or that – once one tries to place the idea in the real, experiential world – then one immediately hits a *reductio ad absurdum.*

The truth is that there are no invisible universes in the Many Worlds hypothesis, or at least none that includes some version of you, because 'you' are there to see them all. I don't mean you, who sees only one strand of the wavefunction, but this 'meta-you' that Many Worlders are forced to invoke but can't really define. *That* 'you', incidentally, must also merge gradually with all possible variants of 'almost-you', right down to the variants that aren't recognizably 'you' at all. So let's be honest – in this view, there is in fact no *you* after all.

To put it another way: no one believes in the Many Worlds

---

* We have been here before too, in a sense. The physicist Pascual Jordan, who worked with Bohr on the Copenhagen interpretation, speculated about a quantum theory of telepathy and clairvoyance, and applied Bohr's notion of 'complementary' quantum states to the psychoanalytic notion of 'split personality' – the coexistence of 'selves'.

interpretation. Those who profess to are unable to give a coherent account of what their belief means, since it implies that the same 'they' denies this belief in another version of reality, that the same 'they' is the only person in the world who disbelieves it, that this 'they' has killed others to 'defend' that belief. They are describing a vision in which miracles, gods and magic exist, where the possibility of science is denied entirely (because of chance exceptions to every physical law). We cannot assign such a vision any ontological status. Far from being the only ones to take the maths seriously, the Many Worlders don't appear to take it seriously enough.

Worlds in collision

All the same, while these are ideas that reach far beyond scientific speculation in the normal sense, they do not deserve scorn purely on that score. They are guesses at questions too taxing for routine investigation to follow: attempts to make sense of what we cannot grasp. Wilczek is right to point out that the historical record shows we constantly underestimate the scope and extent of nature; the discoveries of dark matter and dark energy testify to that and cannot but lead us to suspect that we are still missing some immensely important parts of the puzzle of reality. These parts are not yet visible.

But history also teaches us that attempts to bridge that gap, based on what seems like sound reasoning at the time, are almost invariably wrong. The similarities here with late-nineteenth-century visions of unseen worlds – extra dimensions, invisible intelligences, matter as knots of pure energy, atomized constituents of immeasurably small extent – should alert us to the territory we are entering, in which the traditional tropes of invisibility are still informing the pictures we create. They are a reminder that science is constantly resurrecting old dreams in new guises. The 'invisible multiple selves' in the Many Worlds model, for example, are more fantastical than anything Crookes envisaged – simulacra of you and me identical in every aspect of their lives save that, say, once they sneezed when you and I did not. It is inevitable that some, if not all, of these contemporary ideas about the 'hidden universe' will one day appear as quaint and archaic as *The Unseen Universe* and Crookes's psychic force. If our descendants are fair-minded, they won't laugh on that account, but will recognize the well from which such ideas were drawn.

6 All in the Mind

Q: Did you notice anything other than the players?
A: Well, there were some elevators, and Ss painted on the wall. I don't know what the Ss were there for.
Q: Did you notice *anyone* other than the players?
A: No.
Q: Did you notice a gorilla?
A: A what?!?

Christopher F. Chabris & Daniel J. Simons
The Invisible Gorilla (2010)

Sue had said people were *made* invisible by the failure of those around them to notice them.

Christopher Priest
The Glamour (1984)

In describing how to become invisible in *Dogme et Rituel*, Eliphas Lévi dispensed with the elaborate magical prescriptions of the Middle Ages. It was really a matter of the mind, he said. The secret of invisibility

> is therefore entirely in a power that one may define: that of turning or paralysing the attention, so that light arrives at the visual organ without exciting the regard of the soul. To exercise this power, one must have a will habituated to acts both energetic and sudden, a great presence of spirit, and a no less great skill as causing distractions in the crowd . . . Let a man, for example, who is being pursued by his intending murderers, dart into a side street, return immediately, and

advance with perfect calmness towards his pursuers, or let him mix with them and seem intent on the chase, and he will certainly make himself invisible.

Lévi recounted the story of a priest who was being hunted in the French Revolution by a crowd determined to hang him. He darted down a side street,

> assumed a stooping gait, and leaned against a corner with an intensely preoccupied expression; the crowd of his enemies swept past; not one saw him, or rather, it never struck anyone to recognize him: it was so unlikely to be he! The person who desires to be seen always makes himself observed, but he who would remain unnoticed effaces himself and disappears.

This, said Lévi, shows that 'the true Ring of Gyges is the will'. Talismans such as rings are merely symbolic, a means of focusing the mental energies: 'the symbolic sense of the ring is, that to exert the total power of which ocular fascination is one of the most difficult proofs to give, one must possess the whole science and knowledge of its usage.'

The 'socially assumed' invisibility that Lévi records is something we might all have occasion to desire, whether sitting in the front row of a stand-up comedian's show or being threatened with physical or verbal assault. In history and myth, there is much ambiguity about where 'keeping a low profile' merges with magical powers of concealment – not least, in the idea that Jesus could make himself disappear.

This notion stems from a passage in the Gospel of John describing how Jesus was threatened by the Pharisees (8:59):

> Then took they up stones to cast at him: but Jesus hid himself, and went out of the temple, going through the midst of them, and so passed by.

Did Jesus just blend in with the crowd, or was this genuine magical-miraculous invisibility? While theological interpretations vary, they generally fail to recognize that, in ancient times, skill at the former was not readily distinguished from the latter. In any event, early

Christians found it expedient to maintain social invisibility, as have non-conformist and illegal organizations throughout time, from Rosicrucians to terrorist cells. How is it done?

World of shadows

When Lévi went to England in 1853 he met the novelist Edward Bulwer-Lytton, author of the Rosicrucian fable *Zanoni* (1842) and creator of the fictional force *vril* (see page 43).* The writer claimed to have been enrolled in a Rosicrucian college in Frankfurt and he believed that his occult knowledge gave him the power of invisibility. His grandson's account of how Bulwer-Lytton exercised this power gives us a bathetic picture of what it comprised:

> He would pass through a room full of visitors in the morning, arrayed in a dressing-gown, believing himself to be invisible, and then appear later in the day very carefully and elaborately dressed, and greet his guests as if meeting them for the first time.

The belief in 'willed invisibility' among latter-day occultists took some curious forms. Influenced by Bulwer-Lytton's ghost story 'The House and the Brain', as well as by Bram Stoker's Count Dracula who could pass unseen among men, the American writer Walter B. Gibson created the character of Lamont Cranston, known as The Shadow, who acquired psychic powers in the Orient to 'cloud men's minds' and so, through the use of hypnotism, make himself invisible. The Shadow appeared as a vigilante crime-fighter in a series of novels that were plagiarized by the creators of Batman; one of the first cinematic adaptations of these stories was *The Invisible Avenger* (1958).

In the 1930s and 40s Gibson's output of pulp fiction (sometimes under the pen name Maxwell Grant) was breathtaking: he typically published around twenty novels a year and could churn out 10,000 words a day. He was also a professional magician (and sometime debunker of fraudulent mediums), and wrote many books on magic

---

* His works were immensely popular in their day but have not aged terribly well. The Bulwer-Lytton Fiction Contest annually seeks the worst first line of a novel, for it is he who, in *Paul Clifford* (1830), gave us 'It was a dark and stormy night . . .'

and psychic phenomena, acting as a ghost writer, so to speak, for
Harry Houdini and other illusionists. It's no wonder, then, that the
Far Eastern origins of The Shadow's esoteric knowledge ring true to
the spirit of the times, when Theosophists were insisting that one
should look to the East to find the powers of mind and spirit evinced
in the magic of former ages. 'A modern conjurer who . . . has not
traveled in the Orient cuts but little figure in public estimation,' the
amateur magician Henry Ridgely Evans attested in 1898.

For the Theosophists invisibility was indeed an affair of the mind.
Colonel Henry Olcott gave an account of how Madame Blavatsky
was initiated into the occult arts by the Coptic Egyptian magician
Paulos Metamon:

> From an eye-witness I had it that while H. P. B[lavatsky] was in Cairo
> the most extraordinary phenomena would occur in any room she
> might be sitting in; for example, the table lamp would quit its place
> on one table and pass through the air to another, just as if carried in
> someone's hand; this same mysterious Copt [Metamon] would suddenly
> vanish from the sofa where he was sitting, and many such marvels.
> Miracles no longer, since we have had the scientists prove to us the
> possibility of inhibition of the senses of sight, hearing, touch and smell
> by mere hypnotic suggestion. Undoubtedly this inhibition was provoked
> in the company present, who were made to see the Copt vanish, and
> the lamp moving through space, but not the person whose hand was
> carrying it. It was what H. P. B. called a 'psychological trick', yet all
> the same a fact and one of moment to science.

Olcott described Blavatsky performing the same tricks at home in
Philadelphia, admitting that 'it never occurred to me that I was the
subject of a neat experiment in mental suggestion, and that H. P. B.
had simply inhibited my organs of sight from perceiving her presence,
perhaps within two paces of me in the room'. The Colonel certainly
seems a suggestible sort of fellow.

That invisibility and other illusions are just tricks played on the mind
was proposed even in some old commentaries on magic. The seventeenth-century physician Thomas Ady, whose sceptical view of witchcraft echoed that of Reginald Scot, suggested that much illusionism
relied simply on misdirection – as he put it, a 'busying' of the senses.

Surprisingly, this idea can even be found in the notorious *Malleus Maleficarum*, despite that tract's intention to prove that witchcraft was genuine and satanic. The book states that illusions called glamours or prestiges, wrought by witches, are 'nothing but a certain delusion of the senses' which arise because 'the sight of the eyes is so fettered that things seem to be other than they are'. The issue is *how* such confusions are achieved. This may be done in three ways, the *Malleus* asserts: by devils, by natural magic using the 'virtues' of minerals, and artificially 'by the agility of men who show things and conceal them, as in the case of the tricks of conjurers and ventriloquists'.

It was the devil-mongering that drew the attention of the book's authors, although they were prepared to concede that natural magic did not demand this traffic with demons. What to make of the conjuring tricks is less obvious. A trick might use mechanical devices to make something look like what it is not (such as levitating objects by attaching a fine hair), thus *deceiving* the senses; or it might use something akin to hypnosis or auto-suggestion to *delude* the senses. The Inquisitors seem not to have worried about the distinction, and after all conjurers still employ both techniques today to animate and vanish their props.

Making objects disappear by means of misdirection and prestidigitation has always been among the performing magician's arsenal of tricks, although it is not always clear if the object has vanished (is entirely absent) or is merely no longer visible. As with the 'self-hiding' of small children, what is seen or not in this kind of conjuring is often dictated by the mutuality of gaze between performer and audience. It was said of the great French stage magician Alexander Herrmann that '[i]f his luminous eyes turned in a certain direction, all eyes were compelled (as by some mysterious power) to follow, giving his marvellously dexterous hands the better chance to perform those tricks that were the admiration and wonder of the world'. Disappearances were a central aspect of Herrmann's shows: 'He would place a ring upon the finger of some person, and immediately the ring would vanish from sight.'

It is scarcely any wonder that prestidigitators once aroused awe and suspicion, given what we now know about the capacity of misdirection to make things vanish. Who hasn't failed to spot the gorilla wandering 'invisibly' between the basketball players in that famous video, your mind focused on the pointless enumeration of throws? And who, after that, still believes that invisibility is a kind of magic confined to fairy tales?

The glamour

It became increasingly expected over the course of the nineteenth century that discussions of occult powers such as invisibility be referred to an agency with at least a veneer of scientific respectability, even if that meant Mesmer's animal magnetism or Reichenbach's odic force. But by the end of the century a new 'scientific' explanation for the unseen was emerging, concerned not with hidden forces of the macrocosm but with the preoccupations of the inner world. In the age of Freud, the power of invisibility – like all the psychic powers claimed by the mediums of the day – lurked, as The Shadow might put it, in the minds of men.

This tendency was reflected in the shifting focus of the Society for Psychical Research. At its outset, the society was favoured by physicists such as William Crookes, J. J. Thomson and Oliver Lodge, who sought to rationalize the paranormal in terms of invisible rays and forces. But in the early twentieth century its scientific membership came to be dominated by psychologists and psychiatrists determined to locate invisible worlds not in the ether but in the brain. The two disciplines battled for the right to pronounce judgement. In 1917 the physician Charles Mercier asserted that Oliver Lodge had been duped at séances because, as a physicist, he knew only of 'matter destitute of life, of intelligence, of intention, of volition, of desire, of feeling'. The stage magician John Nevil Maskelyne called the physics laboratory 'quite out of place' when it came to investigating matters of intentional deception – what was needed instead was an expert in misdirection and psychological manipulation (like himself). William Barrett and William Crookes insisted in response that their skill in accurate measurement stood them in good stead for detecting the crude mechanical tricks that fake mediums used. And no one knew better than they how the clumsy intrusion of an observer and his measuring devices might perturb such a 'delicate piece of apparatus' as a true medium.

That Sigmund Freud was a 'corresponding member' of the SPR*

---

* Freud's interest in the occult and paranormal has troubled some of his advocates. His biographer Ernest Jones called it 'one more example of the remarkable fact that highly developed critical powers may co-exist in the same person with an unexpected fund of credulity.' That's obviously a well attested fact, but whether Freud exemplifies it is less clear; it is hard to see how, in the context of his times, he could have failed to be drawn to such things.

was something of an irony, because it was probably Freud who did most to marginalize the society's activities. His theory of the unconscious seemed a more effective, more fruitful and parsimonious way of explaining disembodied voices, sensations of the uncanny, suggestibility and 'possession', requiring no hidden forces but merely our own submerged impulses and desires. He proposed, for example, that belief in ghosts was caused by the repression of disturbing, conflicting feelings of anger and love towards the dead. As the psychotherapist Adam Phillips has put it, with Freud's psychoanalysis 'the supernatural returns as the erotic'.

Yet while investigators of the paranormal seemed intent on bringing these impulses into the open and liberating them, Freud's theory was ultimately geared towards their governance: it was a civilizing theory which, for all its uncomfortable focus on sexuality, was in the end better attuned to the morality of its age. Exotic theories of psychic forces often had a dangerously libertine flavour, which made them seem ever closer to the fringe, as regular psychology on the one hand, and mundane telecommunications technologies on the other, became integrated into society. That the quasi-scientific 'parapsychology' of extra-sensory perception and telekinesis developed by psychologist Joseph Banks Rhine in the United States. Rhine claimed that subjects gifted with 'extra-sensory perception' could guess the symbols on a pack of cards marked with circles, squares, waves and so on with a success rate significantly better than chance: an ability attributed to a so-called psi power, or what American writer Upton Sinclair called 'mental radio'. But who needed this rather hit-and-miss telecommunication when the genuine technologies of invisible forces could reliably deliver on their promise? Radio and television could transmit thoughts and pictures; they were magic that worked. And while parapsychology seemed insistent on revealing the abnormal, ordinary psychology was reassuringly focused on restoring normality.

All the same, if the mind was now the seat of invisibility, Freud implied that it resided in the company of dark and disreputable motives. In *Psychology of Suggestion* (1921) the American psychologist Boris Sidis described an account by a French psychologist named Alfred Binet of an experiment in the hypnosis of an 18-year-old servant girl called Elsie B. The hypnotist put Elsie into a trance and commanded that she not see him when she awoke. And so it was: Elsie was unable to register any action by the hypnotist, even when he stuck pins in her flesh. This, Sidis said, is an example of a 'negative hallucination', a phenomenon first introduced in the 1880s and attested by Freud: not 'seeing things', as in an ordinary hallucination, but 'unseeing things'. But matters went further, as Plato would have foreseen they might. Binet's testimony continued as follows:

Wishing to see, on account of its medico-legal bearing, whether a serious offence might be committed under cover of a negative hallu-cination, I roughly raised her dress and skirt. Although naturally very modest, she allowed this without a blush.

Evidently the habit of older male doctors and professionals to interfere with their younger female subjects was carried over from investiga-tions of the paranormal to studies of the hidden psyche. As writer Roger Clarke puts it, 'in the name of science, professional middle-class men would henceforth lose no opportunity to truss, strap, wire up, restrain and interfere with the flesh and clothing of lower-class female[s] in their purview'. In any event, when she awoke, Elsie did not deny all recollection of the violation, but she spoke of it as though it were a dream. In other words, the unconscious does register the event, the person, the thing unseen – but the conscious mind finds a way to rationalize it away.

This kind of psychological invisibility is explored with great sophis-tication in Christopher Priest's novel *The Glamour* (1984), in which certain individuals are attributed an innate ability to make themselves invisible by act of will, simply by becoming wholly inconspicuous. Those who possess this power give it the name used by the medieval witch-hunters: the glamour. There is, one of these 'invisibles' explains, always someone who you notice last in a room full of people: someone whose unassuming disposition makes them easily overlooked. The invisibles are the extreme cases – those who are, if they choose, *never* noticed. Some may shift in and out of this unseen state; others exist out of sight almost permanently, whether they like it or not. And those miserable wretches drift inexorably to the margins of society, living unkempt in department stores or in empty hotel rooms, stealing to survive. 'As male invisibles grow older', someone with 'the glamour' explains, 'many of them become loners, outcasts even from the outcast society of their fellows, uncaring of how they act . . . the invisibles are a paranoid lot, believing themselves rejects from society, despised, feared and forced into crime.' Their invisibility erodes social norms and degrades moral sensibility.

Priest's novel tells the tale of Richard Grey, a television cameraman who has to reconstruct his life after being badly injured in a terrorist bombing and left with amnesia. He comes to understand that he has been in a relationship with an artist named Susan Kewley, who at first he fails to recognize when she visits him in hospital. He gathers that Kewley is entangled with another man named Niall, who is one of the invisibles. Kewley met Niall because she too has the glamour – and so, Grey gradually understands, does he, which is why he has been able in the past to film dangerous events that others would never have got close to without endangering themselves. Niall is now one of the extreme cases who cannot make himself visible even if he should wish to, and remains unseen even to others with the glamour. As the story elides incompatible narratives, the reader reaches the same understanding as Grey: that the jealous Niall has been shadowing him all along, rewriting his memories, fictionalizing his past. 'The urge to rewrite ourselves as real-seeming fictions is present in us all', says Niall, or the narrator who calls himself that (we're not, in the end, quite sure of anything). 'In the glamour of our wishes we hope that our real selves will not be visible.'

Kewley explains how invisibility came upon her during adolescence. People simply didn't notice her. A car knocked her down on a pedestrian crossing. Her father almost burnt her as she leant against the mantelpiece, when he entered the room and put on the gas fire without realizing she was there. She understood her condition only after meeting a psychic named Mrs Quayle, who told her about Blavatsky and the Theosophists and Aleister Crowley parading in the streets of Mexico believing he could not be seen. Keeping herself visible became an increasing strain for Susan and she began to drift into the demimonde. As a poor art student in London, she discovered that the city is populated by other invisibles when she was chased through Selfridges department store by a middle-aged street dweller, both of them unseen by the thronging shoppers. Then she met Niall.

Priest explores how these illusions of unseeing are sustained, rather than contradicted, by the rational brain. The conscious mind finds explanations, constructs stories, fills in the gaps left by the glamour. Perplexed by Susan's claims of invisibility, Grey visits her parents to see if they will corroborate her story. But they speak only of a 'contented girl, clever at school, popular with the other girls' – a good, considerate daughter. Her mother's account, however, seems oddly barren. She has no anecdotes about her daughter, only 'generalisations and platitudes'. They even speak well of Niall, who they never actually 'saw' at all.

None of this history, Susan explains later, is real. 'I've been invisible to Mum and Dad since I was a kid. The only times they've seen me have been when I've forced myself into visibility.' But that's not what they said, Grey protests. 'That's the way they account for it', Susan replies:

It's how people deal with someone around them who's invisible. They come up with a rational version to explain to themselves what's happened. It's a way of coping.

Priest was influenced by accounts of negative hallucination such as those in physician H. Laurence Shaw's *Hypnosis in Practice* (1977). Shaw explains that a hypnotized subject, told that a person sitting in a chair is invisible, will invent credible reasons not to try to occupy the chair themselves. We demand a cognitively consistent account of what we unsee as well as of what we see. Or to put it another way, we might see only what is consistent with and convenient to our world-view – an idea familiar from the popular and probably apocryphal tales of Western explorers finding their ships to be 'invisible' to 'savages' who had no experience of such things.* Douglas Adams had a character-istically wry perspective on the matter: invisibility in *The Hitchhiker's Guide to the Galaxy* is effected by the SEP field, which can surround objects and render them unseen by making them 'Someone Else's Problem'.

* These stories, variously associated with Columbus, Magellan and James Cook, seem to have originated with the botanist Joseph Banks, who accompanied Cook on the *Endeavour* in 1770. Banks wrote in his diary, while they sailed around the east coast of Australia, that the natives 'seemd to be totaly engag'd in what they were about: the ship passd within a quarter of a mile of them and yet they scarce lifted their eyes from their employment'. This is not, of course, quite the same as experi-encing the ships to be invisible, as even Banks attested: 'I was almost inclind to think that attentive to their business and deafned by the noise of the surf they neither saw nor heard her go past them.' But the myth has stuck.

Grey is induced into negative hallucination by the psychologists who assist his recuperation. 'Stage hypnotists work a similar effect', explains the research student who has just 'vanished' for the hypnotized Grey, 'but they usually make their subjects see people without their clothes on . . . Apparently it works best on members of the opposite sex.' The sexual aspect of invisibility is insistent throughout the novel: at one point Susan leads Grey into a house full of strangers and persuades him to make love unseen, although he can't bring himself to take up her invitation in the lounge in which a group of beery men are watching a football match. Invisible sex has no allure unless it is spiced with voyeurism. Indeed, perhaps invisible sex is *only* voyeurism, as the tale of Gyges and Candaules implied. When the narrator of Nicholson Baker's novel *The Fermata* (1994) acquires an unusual sort of invisibility by being able to arrest time and wander about freely in the frozen instant, he can think of nothing more productive to do with this power than undress women and masturbate. He tells himself that this behaviour is harmless, respectful, almost a necessary and indeed benevolent consequence of his invisibility.

The Glamour is, among other things, a rich compendium of the tropes of invisibility, in which the psychology of unseeing mixes with the rituals of occultism, the spiritual practices of transcendentalism, the tricks of the prestidigitator,* and modernist themes of social alienation that we will encounter in the next chapter. In the best of the 'show don't tell' tradition, Priest builds a persuasive case that these are not different kinds of invisibility but facets of the same phenomenon.

Don't look now

The choice of what to see is not always passive, forced upon us by the magician's misdirection or the hypnotist's auto-suggestion. We are selectively blind to inconvenience just as we are selectively deaf. The question of invisibility is therefore embedded in a broader discourse about sight and vision.

---

* As Priest's later novel *The Prestige* (1995) – better known from the 2006 movie by Christopher Nolan – makes clear, his interest in psychological invisibility (call it glamour, prestige, what you will) is also informed by the techniques of stage magic. The notion of superimposed parallel realities, meanwhile (see Chapter 5), is explored in Priest's recent novel *The Adjacent* (2013).

This was beautifully acknowledged by Hans Christian Andersen in his tale of *The Emperor's New Clothes*. The story is commonly regarded as a parable about the value of defying self-deceiving conventions: the little lad, immune to the peer pressure that ensnares the rest of the townsfolk in a charade of pretending to see what is not there, speaks out honestly about the emperor's nakedness and shatters the mass delusion. There is good reason to believe that Andersen did intend in this way to lampoon the vapid pretensions of the Swedish aristocracy. But as an exploration of the seen and unseen, there are deeper layers to the story.

Some cinematic retellings have lavished most attention on the two swindlers who persuade the emperor and his court that they are making him a set of fine clothes. The garments, they say, are magical and can be seen only by those worthy to do so, not by the inept and stupid. These 'tailors' are really illusionistic tricksters from the tradition of folk magic; yet their trick is not to achieve invisibility by prestidigitation, but to create the somersaulting psychological illusion of denying the invisibility of what is not actually there at all. No wonder we admire their chutzpah.

All the same, they employ precisely the kind of elaborate sensory confusion that stage magicians have always used. If they had simply mimed taking the 'invisible' clothes out of a suitcase, it's doubtful that anyone would have been deceived. To make people see what is not there, just as to make them unsee what *is*, demands a more elaborate performance. The tailors set up looms and enact a skilful dumb-show of cutting and sewing the cloth they have woven. In doing so, says folklorist Maria Tatar, they become more than mere swindlers, for they produce – out of nothing, made of nothing – an object of genuine beauty. 'It is the cloth that captivates us', says Tatar, 'making us do the imaginative work of seeing something beautiful even when it has no material reality.' That is all Andersen's doing, and is why his story-telling is superb. His tailors describe the wondrous patterns and textures of this fabric ('as light as spiderwebs'), so that in fact we – even we, the readers – *do* see it. Then the tale's subject becomes story-telling itself, for it is through narrative, relying on the right word and gesture, that we spin marvels out of thin air and conjure sights that are not real. The fake tailors, Tatar points out, are true artists, whose deceits are sources of delight. Perhaps too, says literary theorist Hollis Robbins, they are Marxist labourers, 'insisting that the value of their labor be recognized ['seen'] apart from its material embodiment.'

How do you feel now about the boy who dispels all this entrancing illusion with crude, materialistic literalism, with a failure to *imagine*? This isn't to say that we have got the tale all wrong – that we should side with the pompous emperor and courtiers and the credulous townsfolk rather than the lad who 'sees through' the deception. It's to say that, as with all the best stories, there is ambiguity, opportunity to sympathize with the fools and villains, temptation to be enchanted by what we suspect we should resist.

All this complicates the question of whether the emperor's new clothes are invisible. If they are not there at all, how can they be invisible? But the tailors have created them in the imagination, so that they are indeed invisible to the emperor and his subjects. Or are we even sure that these folk have not, until the illusion is shattered, colluded to imagine them into visibility? In this way, tales of invisibility may become explorations of seeing and not seeing, of blindness and obscurity and concealment, which are in some respects what myth and fairy tales are all about. 'Arguably enchantment is the core of fairy tales', says folklorist Francisco Vaz da Silva, 'and disappearing into darkness (or becoming blind) is the core of enchantment.'

And who is enchanted by the emperor's new clothes? Not the little boy, but everyone else, for they have overcome the unappetizing sight of the emperor in his underwear (or less) to 'see' the wonderful magic cloth. 'To "see" in a fundamental sense requires overcoming sensory perception', says Vaz da Silva. 'To bar from eyesight the distracting influence of manifest reality [is a] privileged means of grasping the essence of things.' Seers are often physically blind, like Teiresias the prophet of Thebes. Odin became clairvoyant by losing an eye.

Seeing and unseeing are often socially constructed decisions – and to defy their conventions is to destabilize society, for better or worse. That is what the boy does in *The Emperor's New Clothes*, for his cry undermines the deference of the townsfolk towards their monarch.* He is, Robbins points out, a rebel, and Andersen knows it, having the

---

* Andersen's original version, altered just before publication, had no little boy and ended with the townsfolk talking admiringly of the Emperor's 'wonderful new clothes'.

exposed emperor decide that he can preserve his dignity and thus his authority only by continuing the procession rather than by fleeing in embarrassment. But the lad's refusal to 'see' is an inversion of the usual pattern, in which iconoclasts like Joan of Arc and William Blake do indeed see with Odin's more perceptive eye, descrying visions of Christ on the cross or flaming angels in the trees of urban parks.

The novelist China Miéville has explored this theme of unseeing exquisitely in *The City and the City* (2009). Here two fictitious cities called Besźel and Ul Qoma, situated somewhere in eastern Europe or central Asia (the shifting confluence of cultures is part of the point), inhabit the same space, intermeshed in a labyrinth in which one house might be Bész and the neighbour Ul Qoman. The complication is precisely that the cities are *not* mutually invisible, which obliges the inhabitants of each to maintain social etiquette and cohesion by 'unseeing' the other: a skill that requires a lifetime of training and meticulous attention to detail. To fail to observe this convention – or far worse, to step across the invisible but inviolable boundary between one city and the other – is taboo. Offenders incur the intervention of the nebulous, all-powerful and omniscient Breach, whose operatives whisk away the guilty party to some unknown fate. Miéville's novel, on the surface a detective thriller, becomes a meditation not just on political and social segregation in totalitarian states but on how and why we select what we see and what we do not. If we follow Eliphas Lévi by placing invisibility in the mind, we make it subject to the dictates of the will. The power of invisibility is then a question of who commands that will.

7 The People Who Can't Be Seen

I am one of the most irresponsible beings that ever lived.
Irresponsibility is part of my invisibility.

Ralph Ellison
Invisible Man (1947)

The invisible man is a threat to everything we hold dear.

Christopher Priest (2005)
Introduction to *The Invisible Man*

In early December 1897, H. G. Wells received a letter from his admirer Joseph Conrad. Conrad was eight years Wells's senior, but still in the early stages of his career and known simply as an author of exotic romances rather than as an explorer of the dark recesses of the psyche, and he was deferential to his more illustrious friend. The two had begun a correspondence only the previous year, after Wells had reviewed (in appreciative but somewhat lukewarm terms) Conrad's *An Outcast of the Islands*. Their later relationship was sometimes strained by the contrasts in their characters and styles – Conrad's biographer Frederick Karl says that '[f]rom the start, the two were as different as poetry and applied science'. But Conrad's letter in 1897, occasioned by the publication of Wells's new novel, was almost gushing – yet also perceptive:

> I am always powerfully impressed by your work. Impressed is *the* word
> O! Realist of the Fantastic, whether you like it or not. And if you want
> to know what impresses me it is to see how you contrive to give our

humanity into the clutches of the Impossible and yet manage to keep it down (or up) to its humanity, to its flesh, blood, sorrow, folly. *That is the achievement!* In this little book you do it with an appalling completeness.

This description of Wells as the 'Realist of the Fantastic' has been much quoted by his biographers and critics as the perfect distillation of the man's oeuvre. He takes the most fantastical notion and expounds it with the kind of detail more usually applied to the literary social realism of his age: tea and cakes and time machines. It's a tellingly apt phrase that, Conrad's intention notwithstanding, can be used to disparage Wells as much as to praise him: his literalism sometimes threatens to compromise the allegorical potency of his fantasies. But nowhere is the phrase more appropriately applied than to the book that inspired Conrad to coin it – which was, of course, *The Invisible Man*.

One of Wells's most commercially successful novels, *The Invisible Man* presents the first popular, truly scientific recasting of the myths of invisibility. Any hint of occult magic is scrupulously excluded and Wells goes to much trouble to create an impressively plausible account of how science alone might make it happen. This is precisely why the book offers an important view of what invisibility might mean in the modern age: how the mythical attributes are sustained yet at the same time mutated and corrupted by the incursion of science. On the one hand, *The Invisible Man* is the touchstone for all subsequent techno-logical attempts to make invisibility real. On the other hand, by showing how modernity recasts invisibility not as a power but as a blight, it created a template for metaphors of social alienation and impotence.

Strange affairs in Sussex

By demonstrating the corrupting effects of the power of invisibility, *The Invisible Man* is a retelling of Plato's myth of the Ring of Gyges. It isn't clear if this was Wells's explicit aim, but that seems likely. He was deeply impressed by the *Republic* when he read it in his youth, not least for the alternative it offered to the stifling orthodoxy of Victorian society. An advocate of socialism and free love, Wells seemed to feel that Plato's political theory justified both of those attitudes. In that much he was neither the first nor the last to fit the Greek philosopher's ideas to his own preconceptions.

The Invisible Man begins with the arrival, one snowy day in December, of a 'strange man' at the Coach and Horses public house in the fictional village of Iping in Sussex. Given the weather, it is no surprise to the proprietress Mrs Hall that the visitor is wrapped up from head to foot; but when he sits for his meal and takes off his hat, she gets a shock. He wears an odd pair of large blue goggles, and his forehead is entirely covered with a white bandage between which tufts of hair poke out to give him 'the strangest appearance conceivable'. All one can see of his face is a shiny pink nose. 'The poor soul's had an accident or an op'ration or something', Mrs Hall opines to the maid.

The man is terse and irritable to the point of rudeness and when he takes a room at the inn he demands to be left alone. Inevitably, word spreads in the close-knit rural community, and the folk of Iping begin to speculate about who or what the man is. They see peculiar things: one swears that the leg glimpsed through a tear in the trousers was quite black, which, combined with the pink nose, implies the man is a 'piebald', and doubtless hiding it through shame. The village doctor attests that he saw nothing but an empty sleeve where the man's arm should be.

When Mrs Hall confronts her odd guest over an unpaid bill, in fury he exposes the truth, tearing off his clothes to show that he is invisible. Mrs Hall flees in terror. The village constable, Jaffers, arrives to arrest the man, now a seemingly headless figure in a coat eating alone in the pub, but he strips off again and escapes, and the Invisible Man becomes a fugitive.

We discover that he is a scientist named Griffin, a former medical student of University College in London who, fascinated by light, took up physics instead. He became convinced that he knew how to make substances invisible, but after leaving the university and working as a professor's assistant in a provincial college, he refused to share his secret with his boss, fearful that credit would be stolen from him. Eventually he left the college and, with money stolen from his father, returned to London where 'in a big ill-managed lodging-house in a slum' he experimented to find the chemical formula of his potion of invisibility. 'To do such a thing would be to transcend magic', he says.

'And I beheld, unclouded by doubt, a magnificent vision of all that invisibility might mean to a man, – the mystery, the power, the freedom.'

Griffin's trial on a cat succeeds in making it vanish, but when his landlord, suspicious of his secrecy, tries to expel him from his room, Griffin swallows the potion, smashes his equipment, sets fire to the house and slips invisibly onto the London streets.

Whilst on the run from the Coach and Horses, Griffin tells all this to a former colleague from University College, a doctor named Kemp who is now a respected scientist living in the port village of Burdock, near Iping. By this stage Griffin has been terrorizing Iping and its environs, unwillingly assisted by a tramp named Marvel who he has impressed into his service through dire threats. He has now descended into sheer, crazed malice, determined to subjugate the world (starting with rural Sussex) in a Reign of Terror. Escaping from Kemp's house after the doctor has surreptitiously called the police, he murders a passing old man by bludgeoning him with an iron bar, for no other reason than to demonstrate his power and remorseless intent. He leaves a note at Kemp's house:

> This announces the first day of the Terror. Port Burdock is no longer under the Queen, tell your Colonel of Police, and the rest of them; it is under me – the Terror! This is day one of year one of the new epoch, – the Epoch of the Invisible Man. I am Invisible Man the First.

To mark the advent of this new epoch Griffin will take revenge on Kemp by making him the first victim: 'He may lock himself away, hide himself away, get guards about him, put on armour if he likes; Death, the unseen Death, is coming.' But this declaration shows the police and citizens where to focus their manhunt, and as Griffin chases the fleeing doctor into the town of Port Burdock, a mob is able to apprehend him, whereupon they beat him to death in a terrified frenzy. In death Griffin returns to visibility: 'naked and pitiful on the ground, the bruised and broken body of a young man about thirty'. And 'there ended the strange experiment of the Invisible Man'.

Or not quite. For Marvel has managed to steal off with Griffin's notes, which contain his secret equations and formulae. The former tramp becomes wealthy by telling his tale in music halls and opens a pub called the Invisible Man – and here, every Sunday morning, he pores over the incomprehensible notes, hoping beyond hope that one day he will decode them. 'Full of secrets', he mutters to himself, 'Wonderful secrets!' But, Marvel assures himself, if he ever works out the secret, 'I wouldn't do what *he* did; I'd just – well!'

One aspect of invisibility that Wells captures with almost gleeful energy is the revulsion that it can engender – the sense of impropriety and dislocation occasioned by viewing an absence where a body should be. We do not need to be a superstitious bumpkin to share Mrs Hall's horror at the 'vast and incredible mouth' that she glimpses as Griffin dozes in his chair, which seems to have 'swallowed the whole of the lower portion of his face'. When her husband goes to Griffin's aid after he has been bitten by a dog and catches him unawares in his room, he sees 'a waving of indecipherable shapes', an impression of which 'his vocabulary was altogether too limited to express': he can't even say what it is he has (or has not) seen. The residents of Iping, seeing Griffin disrobe at last, are horrified: 'They were prepared for scars, disfigurements, tangible horrors, but *nothing!*'

Accompanying the fright of *not* seeing what we should is the paranoid fear of being the prey, the object of voyeuristic malice, of something pervasive but unseen: 'He may be watching me now', Kemp avers uneasily. It is the disembodied voice of his persecutor that first fills Marvel with fright. What is heard but not seen carries a frisson long evoked by the supernatural. It is one thing to say 'be not afeard', but when the isle is full of noises we can scarcely suppress our dread, even if they claim they will 'hurt not'.

Inventing invisibility

On a superficial reading, *The Invisible Man* might seem a mad-scientist tale not unlike *Frankenstein*: a lone genius discovers a potent and dangerous secret, is driven crazy by it and ultimately comes to a bad end. There are undoubtedly elements of Mary Shelley's cautionary fable in Wells's novel, not least in casting Griffin as someone who cuts himself off from the social intercourse that might otherwise have rescued him from his fateful obsession. But there is much more to it than that (as there is in *Frankenstein* too). This is no moral fable about scientific hubris. Wells, after all, was not only highly scientifically literate – he wrote popular science and counted the eminent biologist Julian Huxley among his friends – but also considered the advancement of science a positive force for shaping a utopian society. This didn't hold him back from depicting science abused – *The Island of Dr Moreau* is the most obvious example – but we can't cast him as the reactionary romantic fretting that scientists will abuse their power.

Wells's knowledge of and enthusiasm for science is clear from the effort he makes to provide a scientific explanation for Griffin's invisibility. There can be no more magic rings, spells or amulets; now, modern physics, chemistry and biology are enlisted to supply a detailed and almost convincing prescription. One of the novel's reviewers attested that the reader 'is really almost persuaded that one's own ignorance of the true meaning of scientific formulae alone prevents a full apprehension of the process by which Griffin is able . . . at last to fade away himself out of human sight'. With reference to 'Röntgen Rays and other still mysterious vibrations', the reviewer adds, Wells can 'reduce the impossible into terms of the probable.'

This was Wells's objective in all of his early fantastical 'scientific romances', such as *The Time Machine* and *Dr Moreau*. He laid out his agenda in the preface to a 1934 anthology of these tales:

> For the writer of fantastic stories to help the reader to play the game properly, he must help him in every possible unobtrusive way to *domesticate* the impossible hypothesis. He must trick him into an unwary concession to some plausible assumption and get on with his story while the illusion holds . . . Hitherto, except in exploration fantasies, the fantastic element was brought in by magic. Frankenstein even, used some jiggery-pokery magic to animate his artificial monster . . .*
> But by the end of the last century it had become difficult to squeeze even a momentary belief out of magic any longer. It occurred to me that instead of the usual interview with the devil or a magician, an

---

* Wells is a little unfair to Mary Shelley, whose tale of reanimation is striking specifically because it is secular and non-supernatural. She is vague about Frankenstein's experiments, but one is left in no doubt that they are meant to be in accord with the scientific understanding of her times.

ingenious use of scientific patter might with advantage be substituted. That was no great discovery. I simply brought the fetish stuff up to date, and made it as near actual theory as possible.

This was an innovative aspiration in Wells's time, and undoubtedly contributes to the enjoyment one can find in his books. Science fiction that is scientifically plausible has thrived ever since, notably that of well-informed authors like Isaac Asimov and Arthur C. Clarke. But Wells puts his finger on the salient matter almost inadvertently: 'to domesticate the impossible'. There is a sense of mundanity in this formula. That is partly the point, as we shall see: the almost tedious practical concerns and constraints of invisibility are central to Wells's story. However, his remarks here expose the misrepresentations that can accompany attempts to literalize myth as science. The whole point of mythical (and folkloric) magic, as opposed to the natural or demonic practical magic of the Middle Ages, is that *it doesn't matter how it is done*. It requires no special skills or lengthy procedures, no spells or incantations: it just *happens*, provided that one has the right power or talisman. That is why folktales rarely feature mighty wizards or supernatural fairies: the magic is wrought by old women, animals, craftspeople, or an object discovered by chance. Most importantly, it is not especially miraculous. In the closed logic of the folktale, magic is normal: people use it, sometimes they suffer for it or are fooled by it, but they are never very surprised by it.

In trying to turn mythical magic into science, Wells was therefore confusing categories: he was attempting to convert a symbolic power into a physical effect. The occultists and natural magicians made the same mistake, which is why they too created convoluted and unreliable recipes for realizing such mythical properties as invisibility. Compare Plato's Ring of Gyges with those 'prepared' by a medieval prescription as seen in Chapter 2. Glaucon is utterly unconcerned about how the ring came to have its power, or how it was forged, whereas for the occult magician everything depends on these things. Wells's literalizing of invisibility compromises its powers and for that reason compromises its symbolic role. (But not entirely – for as we will see, Wells could not in the end wholly relinquish mythic invisibility.) This confusion of roles continues to complicate today's alleged technologies of invisibility.

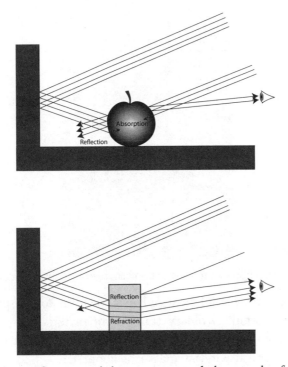

An opaque object reflects some light to our eye, and obstructs that from the background (top). A transparent object reflects a little light but transmits some from the background, at the same time refracting it so that the background is distorted (bottom).

So how, then, does the Invisible Man vanish? It's a matter of optics. On the whole, we see objects because they absorb and reflect light. An object's colour results from light absorption, which removes certain wavelengths from the light that is reflected to our eyes. When sunlight falls on a rosy apple, the blues and greens are absorbed and the reds are reflected. These processes of absorption and reflection also obscure what lies behind the object, because they prevent light from that part of the background from reaching our eyes.

Fully transparent objects such as glass don't absorb visible light: it can pass right through, and the light coming from behind can reach our eyes. But a glass object such as a wine glass isn't invisible. Partly that's because some light is reflected from the smooth surface, so the outline is picked out in the bright reflections. But partly it is because the path of light is altered as it travels from air into glass: because the speed of the light is slowed, the light ray is bent. The amount of slowing and bending is measured by the object's refractive index,

Reflection and refraction by a transparent object prevent it from being truly invisible.

which is the ratio of light's speed in a vacuum to its speed in the material in question. Ordinary transparent materials have refractive indices greater than 1 (the refractive index of air): that of water is 1.33, and for typical window glass it is about 1.5.

This bending of light caused by refraction distorts the image of the background that is visible through a transparent object, so that we can scarcely fail to see the boundaries and contours of the object itself. For a thin sheet of window glass there is very little refraction, however, and if we happen to come at it from an angle that doesn't catch strong reflections either, then it might indeed be unseen and we risk bashing our noses into it.

By matching the refractive index of oil with that of a glass rod dipped into it, the glass rod can be made to 'disappear'.

Suppose that a way could be found to make the human body as transparent as glass, and also to eliminate reflections from its surface. It would then be pretty hard to see at a glance. But for it to be perfectly invisible, you must eliminate refraction as well. For glass, this can be done by immersing it in a transparent liquid with the same refractive index. Then there is no deviation of light rays as they pass from the liquid into the material. Glass objects already become harder to see if you put them in water, since they have similar refractive indices. In pure alcohol (ethanol: refractive index 1.36) glass becomes still more invisible, and in clear baby oil or liquid benzene (refractive index 1.5) it vanishes.

So there you have a prescription for scientific invisibility: make flesh transparent, with a refractive index as close to 1 (that is, to that of air) as possible. Can it be done? In water rather than air, nature comes close already: the bodies of some sea creatures such as jellyfish are more or less transparent, and hard for predators to make out in the gloomy depths.

All this Griffin explains to Kemp, right down to the fact that a sheet of glass 'in some denser liquid than water . . . would vanish almost altogether . . . a transparent thing becomes invisible if it is put in any medium of almost the same refractive index'. But a man is not transparent, Kemp protests. On the contrary, says Griffin, 'the whole fabric of man except the red of his blood and the black pigment of hair, are all made up of transparent, colourless tissue'. This isn't quite true – we have skin pigments, for example – but it is closer to the truth than we might imagine, and closer still for Griffin, who is an albino.

'And suddenly, not by design but by accident, I made a discovery in physiology', Griffin explains. 'You know the red colouring matter of blood; it can be made white – colourless – and remain with all the functions it has now!' To achieve this, Griffin devised a drug (the same stuff whose effects he later seeks to reverse by chemical experimentation) that removed the light-absorbing pigmentation of blood. But Wells knew that this would at best give a body the appearance of glass. How to lower the refractive index until it matches that of air? No physical process was known that would do this, so Wells did after all need a little magic – not the old-fashioned alchemical sort, but the modern invisibility magic of Maxwell and Röntgen. Griffin uses dynamos and oscillators to generate 'a sort of ethereal vibration' – not X-rays but another kind hitherto unknown – to which the material was exposed.

Even if in the end Wells needed this sleight-of-hand pseudo-science, the argument is impressively made. It is certainly more sophisticated than that in an earlier attempt to invoke science fictional invisibility, a short story called 'The Crystal Man' written in 1881 by the American Edward Page Mitchell. Mitchell's scientist, experimenting in a chemistry lab in Germany, finds it sufficient merely to eliminate all pigments from the body, imagining that transparency alone is equivalent to invisibility. Whether or not Mitchell's tale was known to Wells, it is a threadbare precursor, a fleeting yarn in which the invisible scientist commits suicide when spurned by his lover.

Jack London made the opposite mistake in his later short story about invisible men, 'The Shadow and the Flash' (1903), in which an ingenious chemist named Lloyd Inwood asserts that an object need only be perfectly black, absorbing all light that falls on it, for it to become impossible to see. 'With the right pigments, properly compounded', he says, 'an absolutely black paint could be produced which would render invisible whatever it was applied to.' Inwood doesn't seem to reckon on how light from behind the object is meant to reach the eye. (What he is actually making here is more akin to the kind of stealth coating used to hide aircraft from radar: see page 251.)

London has some inkling of this problem, however, for Inwood's rival Paul Tichlorne dismisses the claim by saying that the 'invisible' object will cast a shadow. Tichlorne, meanwhile, is convinced like Mitchell that transparency is the key to becoming unseen. But he discovers that the dog he makes transparent by chemical means produces occasional rainbow flashes, like the atmospheric phenomenon of sun dogs caused by refractive effects. Locked in bitter competition, the invisible Inwood and Tichlorne finally fight to the death in a mad whirl of shadows and flashes, light against dark, neither man ultimately able to defeat the compromising physics of optics.

Wells was too much the scientist to believe that his account was flawless either. In particular, he knew that vision is possible only because pigments in the retina absorb light and convert the energy to nerve signals. If the retinae are fully invisible, their owner is blind. What's more, the optics of vision relies on refraction: the eye produces an image by acting as a lens. Wells admitted as much in a letter of 1897 to the writer Arnold Bennett and he allowed himself to hint at it in the novel. The cat on which Griffin first tries out his procedure is rendered invisible except for the 'two little ghosts of her eyes': an optical Cheshire cat. And when Griffin takes the drug, he sees in the mirror 'nothing save where an attenuated pigment still remained behind the retina of my eyes'.

This seems like a minor point that Wells was right to brush aside for the sake of his story. But it is crucial to the meaning of Wellsian invisibility. When the folkloric hero dons a cap of invisibility, he remains perfectly able not only to move and speak but to see, for otherwise the power is all but useless and there can be no narrative through which to explore invisibility as a moral or social construct. Because Wells wishes to use invisibility that way too, he is forced knowingly to utilize this folkloric invisibility, however hedged by scientific and practical constraints. The result is a somewhat untidy mixture of modes: realist and fabulous. It doesn't prevent Wells from spinning a good yarn – but as he doubtless realized, it means that he is not really exploring the consequences of scientific possibilities, but is using science as cover to carry his modern readers along with him. And that, argues literary scholar Tatyana Chernysheva, is often how it goes in science fiction. It is why science fiction should be read (and, I believe, written) not as a prediction about where science will take us but as the imaginative use of science to accentuate the contours of – in order to explore more searchingly – social and technological questions that are already upon us.

The problem with invisibility

The troublesome features of the scientific theory of invisibility are as nothing compared with the inconveniences of the practice. Wells's plot rests on the contrast between the grandiosity of Griffin's dreams and the banal realities that hinder him. As he escapes from his London hovel once he has made himself invisible, he tells Kemp, '[m]y head was already teeming with plans of all the wild and wonderful things I had now impunity to do'. But at once he finds that he can scarcely even walk, since being unable to see his legs and feet makes this an unexpectedly difficult operation. He stumbles out onto the street, excited at how he can 'revel in my extraordinary advantage', only to be knocked about by people bumping into him. Far from being a god amongst men, he is literally nothing to them. In no time he is left shivering in the 'thin slime of mud that covered the road', naked and bruised and freezing in the January air. Dogs sense his presence and yelp, his muddy footsteps are clear to see, and as snow starts to drift down he realizes that it will betray him if it settles on his body.

One might have expected a man of Griffin's intellect to have taken the precaution of making clothes invisible too, not least because his very first successful experiment is with a piece of 'white wool fabric'. The oversight is intentional, because so many of the inconveniences needed for the narrative flow from it, but that doesn't make it any the more plausible. Yet part of the effect that Wells wishes to conjure in the early part of the book relies on the strangeness of seeing an invisible man in visible, hollow clothes. For the more farcical aspects of that situation Wells admitted that he took his inspiration from a satirical poem written by W. S. Gilbert, best known as Arthur Sullivan's librettist, under the pen name Bab. Called 'The Perils of Invisibility', it told of a hen-pecked man who is granted the gift of invisibility to escape from his nagging wife. She contrives to ensure that his clothes remain visible, so that he is forced to walk the countryside as an 'empty suit', terrifying everyone he meets. The poem ends with a comical foreshadowing of Griffin's exploits in the lanes of Sussex:

> At night, when all around is still
> You'll find him pounding up a hill;
> And shrieking peasants whom he meets,
> Fall down in terror on the peats!

Even when the naked Griffin, cold and desperate on London's streets, takes refuge in a department store, where he should be able to help himself to whatever he likes, he takes little solace from it. He can't keep the clothes he filches, since once he is clothed the shop staff can see him and give chase when they arrive in the morning. Even eating is risky, because food remains visible for a time in his digestive system until it is assimilated. (Don't think too hard about that.) 'I went over the heads of the things a man reckons desirable', he says. 'No doubt invisibility made it possible to get them, but it made it impossible to enjoy them when they are got.' His most urgent priority, he soon realizes, is to undo his invisibility – if not literally, then by getting hold of some costume that would at least turn him into a 'credible figure' and not some 'strange and terrible thing'.

Literary critic Paul Cantor rather ingeniously finds in all of this a critique of capitalism and consumerism from the socialist Wells: the apparent opulence of the department store fails to offer Griffin any real comfort. Cantor even draws a parallel between Griffin and Adam Smith's Invisible Hand, the self-organizing agency that stabilizes capitalist societies, for it is soon Griffin's invisible hand that is lifting money out of people's pockets and spiriting it away, to be carried (for Griffin himself can't keep it, visibly, on his person) by the hapless Marvel. It's a nice idea, but if Wells intended this subtext then, as Cantor shows, he doesn't develop it in any consistent or convincing way.

More plausible, and more significant for the twentieth-century narratives of invisibility, is the suggestion that Griffin represents a challenge to bourgeois middle-class complacency: a perfectly likely target for the libertarian and left-wing Wells. He stresses that Griffin is a socially disadvantaged outsider – as Wells was himself, the son of a poor shopkeeper and a lady's maid – and the early part of the book invites the reader to revel in the anarchy that the Invisible Man wreaks in society. Which is all very well, except that the society that Griffin assaults is not middle-class London but bucolic Sussex. Readers' sympathies apparently vary, but I can't help seeing something perverse in enjoying the spectacle of the thoroughly unpleasant Griffin raining blows on a few terrified villagers.

It's this aspect of Griffin's character that defeats any aspiration the book might have had to parallel the moral story of Gyges. There's no reason to suspect that at the outset Gyges was not a decent fellow; of Griffin, in contrast, we never see anything remotely likeable. 'He has cut himself off from his own kind', says Kemp – but this is the case even before he has made his discovery. Most deplorable of all is the way he steals the money for his experiments from his ageing father, who was already in debt and ends up shooting himself in disgrace. 'I did not lift a finger to save his character', Griffin admits. 'He seemed to me to be the victim of his own foolish sentimentality.' He goes to the funeral out of social obligation, 'but it was really not my affair'.

For all the frequent suggestions that *The Invisible Man* is a Faustian tale, then, the fact is that Griffin has no pedestal from which to fall. He is a monster from the start, and when he begins to rave about his dreams of tyranny we are hardly surprised. His Jekyll is already too close to his Hyde for one to draw any moral about how the irresponsible pursuit of knowledge can corrupt.* Some critics seem strangely determined to enlist our sympathies for Griffin – John Batchelor, for example, claims that the reader is disappointed when, towards the end of the book, 'Wells turns against the implications of his invention and bends the knee to a conventional morality in which he clearly has no real faith'. Personally I root for Mrs Hall and old Constable Jaffers all along.

So what *was* Wells trying to tell us? It's fair to imagine that this proselytizer for science may have wanted his audience to conclude something favourable about its methods and virtues, and at the very least he suggests that, faced with the inexplicable, we won't get far by responding with the kind of superstitious ignorance displayed by the Iping villagers. Today the fashion among scientists and their advocates is to demand evidence-based reasoning for everything. But in the late nineteenth century scientists were happy to admit that the data itself is not always adequate for reaching definite conclusions, and that to reason beyond them one might need to add another ingredient: imagination.

It is via the imaginative leap, argued John Tyndall in his essay 'Scientific Use of the Imagination' (1870), that 'we can lighten the darkness which surrounds the world of our senses'.† As we saw, the science of Tyndall's day – and its physics in particular – was becoming increasingly concerned with the invisible, with particles and rays that the eye could not perceive. Only by admitting the hypotheses of atoms, ether and fields could scientists make sense of the phenomena they observed. Literary critic Steven McLean argues that '[l]inked together by the scientific imagination, the sensory clues which provide the basis for the wrongheaded guesses concerning [Griffin's] identity . . . could have successfully anticipated the disguise later discarded by the stranger'.

Perhaps so. There certainly seems to be something in McLean's suggestion that, given the context of Tyndall's discussion and the directions of late nineteenth-century physics, it was only in the 1890s that *The Invisible*

---

* Robert Louis Stevenson's *The Strange Case of Dr Jekyll and Mr Hyde* (1886) surely provided Wells with some inspiration. For example, when the fleeing Griffin callously knocks aside a child and breaks its ankle, the scene recalls the way Hyde 'trampled calmly over the child's body and left her screaming on the ground'. And Griffin's bludgeoning murder of the old rambler Mr Wickstead with an iron rod echoes an assault perpetrated by Hyde with his heavy cane.

† I will only point out in passing the etymological link between 'imagination' and 'magic'.

Man could have been written. But there are too many contradictions and flaws in the book to take from it any clear allegorical message. It is largely because of these shortcomings, rather than by the author's design, that interpretations of Wells's intentions vary so widely.

The book's greatest significance – I am not sure this was intended – resides in what it does to the myth of invisibility. It's not meant as criticism when I say that Wells domesticates it – that much, at least, was the author's avowed intent. But he may not have seen quite how much it was diminished as a result. That the power of invisibility should become a curse is not a wholly new notion; but with Wells, both power and curse become pathetically constrained. Griffin dreams of world domination but manages little more than to scare a tramp and murder an old man. His demise has no real tragedy in it, but is like the lynching of a common criminal, betrayed by sneezes, sore feet and his digestive tract. And the magic itself is a compromised affair of refractions and albinism, barely able to hide a cat. In all this, Wells shows us what to expect from science.

The invisible man at the movies

Whatever the imperfections of Wells's novel, he can hardly be held responsible for the indignities to which his concept was subjected in

Griffin confronts the villagers of Iping in James Whale's film *The Invisible Man* (1933).

the following decades. James Whale's 1933 film, with Claude Rains in the title role, follows the book's plot to a considerable degree and yet manages to subvert its messages. Not only is Griffin totally deranged throughout, given to outbursts of melodramatic cackling, but we are not even permitted to attribute this mental instability to any *psychological* consequence of his powers or condition. Instead, the cause is pharmacological. Part of the process of becoming invisible now involves ingesting a herb called Monocaine that grows in – where else? – the East and which the hyper-rational Kemp has since found to induce mental disorder.

Wells did what he could to avoid the worst distortions of his book. When he sold the film rights to Universal Studios, he insisted on power of veto on the screenplay. This made for a lengthy and difficult gestation for the project, which involved four directors and ten screenwriters. Wells rejected one screenplay after another, including one from Whale himself, until finally approving one by the writer and journalist John Weld that, in a boldly radical departure for Hollywood, actually took the book as its source. Better still, the scriptwriter employed for the job was Wells's friend Robert Cedric Sherriff, an Englishman like Whale himself. Sherriff later claimed that Wells had no problem with the idea of making Griffin an 'invisible lunatic' – it would, the author acknowledged, 'make people sit up in the cinema more quickly than a sane man' – but he didn't like the idea of making Griffin's madness drug-induced. He told Sherriff that it ought to be the condition of invisibility itself that drives Griffin insane. 'Obviously', wrote Sherriff, 'it was a better idea from an artistic point of view, but I did not think it practical for the screen.' How could one show Griffin's gradual mental deterioration when his face was covered in bandages?

Wells continued belatedly to protest this point with Whale at the London premiere, saying '[i]f the man had remained sane, we should have had the inherent monstrosity of an ordinary man in this extraordinary position'. Whale – whose previous smash hit *Frankenstein* (1931) had similarly undermined the moral message of the creature's violence by making Frankenstein inadvertently give him the brain of a deranged criminal – replied that '[i]f a man said to you that he was about to make himself invisible, wouldn't you think he was crazy already?' Wells had the grace to laugh at that, and in his 1934 biography he praised the film as 'excellent'. One supposes he had seen more than enough of Hollywood by then to know how much worse its fate might have been. Giving Griffin a spurious love interest, played by Gloria Stuart, was not much to complain about in comparison.

Stuart later wrote wryly about Rains's efforts to upstage her. Although this was his first film role (he admitted to having previously only watched about six movies in his life), Rains was an experienced theatre performer who knew how to chew up the scenery. The fact that his face was not seen until the final frames (there are no flashbacks to Griffin's pre-invisibility experiments) must only have sharpened his determination to make an impression. That lack of on-screen exposure may have played a part in Universal's failure to secure a big star for the role – who wanted to be invisible in Hollywood? Boris Karloff and Colin Clive (Whale's Dr Frankenstein) declined and Rains himself didn't realize he would play the part with bandages covering his features throughout, since he'd not considered it a wise precaution to read Wells's book before accepting the role.

Universal saw *The Invisible Man* simply as a horror film in the lucrative tradition of *Frankenstein* and *Dracula*, albeit with some comedic aspects. Yet Whale's choice of subject is another indication of the alliance of early cinema with the long tradition of optical magic, and he evidently relished the chance to show audiences what 'real invisibility' would look like. Objects are carried and thrown by invisible hands, Rains's body appears as headless as the decapitated men in the tricks of medieval mountebanks, and his trousers dance with disembodied lunacy down a country road singing 'Here we go gathering nuts in May', an image straight out of W. S. Gilbert's whimsy.

For these 'empty clothing' shots, Whale made ambitious use of the 'matte' technique in which shots of the actor are superimposed on the scenic background, with Rains' body obscured by tight black velvet clothes and mask that vanish in the overlay. Other effects were achieved with mechanical rather than photographic trickery: the bicycle that goes speeding down a village street by itself was operated by wires strung to an overhead cable. But these effects were sometimes wanting, and tens of thousands of individual frames of footage were touched up with dyes painted on by artists using tiny brushes and looking through a microscope, a feat of concentration that couldn't be sustained for more than two hours at a time. Although these special effects are impressively realized, few were entirely new. The animated and dancing clothes were anticipated, for example, in Méliès' ghost story *The Bewitched Inn* (1897) and Cecil Hepworth's *Invisibility* (1909) – the latter a Gilbertian comedy in which a man buys a powder of invisibility to evade his nagging wife. The riderless bicycle comes straight from *The Invisible Thief*, made by the Gaumont film company in France in 1910.

The risk was that actors playing to an absent performer would appear to be doing just that. Whale found it necessary to devise 'bits of "business"' for the scenes in which Rains was wholly invisible, to let the audience know he was 'there' – for example, having him sit in a rocking chair which could be moved back and forth. Media scholar Keith Williams points out that these illusions are actually examples of metonymy, in which a physical object stands in for the person associated with them (such as when kingship is called 'the crown'). In Whale's movie, Griffin's dressing gown and other clothes, even stolen money,* 'become' the man himself.

This business of signalling the presence of the invisible was already written into Wells's novel, and even there it was anticipated by other authors. Only four years before *The Invisible Man* was published, Ambrose Pierce depicted the visible impact of the invisible in his short story 'The Damned Thing', in which a game hunter is killed by a creature that cannot be seen because it is of a colour outside the range of human vision. As the victim's companion relates,

> I observed the wild oats near the place of the disturbance moving in the most inexplicable way. I can hardly describe it. It seemed as if stirred by a streak of wind, which not only bent it, but pressed it down—crushed it so that it did not rise.

A still more explicit foreshadowing of Wells's filmic metonymy may be found in the ghost story 'No. 11 Welham Square' (1885) by Herbert Stephen, in which a seated ghost is 'seen' by the effect of its weight on the chair:

> As I looked at this chair it struck me that the seat was considerably depressed, as though some one had recently sat down upon it, and the

---

* The floating money, accompanied by Griffin repeatedly barking the word 'Money!', offered a visual pun on the old expression that money talks.

seat had failed to resume its ordinary level. This surprised me, for I had sat in the chair that morning and felt sure the springs had then been in good order. I looked at the antimacassar. Towards the top it was pushed up in wrinkles. As I looked, it occurred to me that it was impossible for it to hang in such a manner by itself. It looked for all the world as if an invisible but substantial human frame was then actually sitting in the chair.

Notice that this is a strangely substantial ghost; all the more so when Stephen's hero wrestles with it at the end of the tale. It is, indeed, not a traditional ghost at all, but rather, like Fitz-James O'Brien's unseen demon (page 85), an invisible man of the kind that writers such as Mitchell and Wells were beginning to imagine.

Return of the invisible man

Others took far greater liberties with Wells's concept than Whale did. But even the worst of them could be revealing about attitudes to and meanings of the unseen. When *The Invisible Man* became, inevitably, *The Invisible Woman* (1940) – this time played solely for laughs rather than chills – the film censors were confronted with a metaphysical dilemma for which a career immersed in Hollywood's confections had not prepared them. Is it decent to show a naked woman on the screen, *if she is invisible?*

That frisson, and the whole notion of invisible sex, has proved irresistibly titillating to movie moguls. The special power of the Invisible Girl Sue Storm Richards,* one of the Fantastic Four, might seem innocent enough, until it requires Jessica Alba, in the 2005 film of the superhero quartet, to invisibly occupy, and then vacate, her underwear. A similar striptease was even more tawdry in the risible *The Invisible Kid* (1988), but even that seems innocent compared with the sexual licence of Paul Verhoeven's 'updating' of Wells's story *Hollow Man* (2000). The best that can be said of the scene in which an invisible Kevin Bacon undresses and then rapes the sleeping

---

* The Invisible Girl owed her powers to invisible rays, in this case cosmic rays to which she was exposed on a space mission. Another 'invisible force' enables her to turn other objects invisible.

Elisabeth Shue is that Plato might have murmured 'I told you so'. All the same, it is in its very lack of subtlety that the blockbuster movie so often exposes the undercurrents of myth. If Plato had worked in Hollywood, he might have summed up Glaucon's story with *Hollow Man*'s tagline: 'What would you do if you couldn't be seen?'

The well of cinematic inspiration, if it can be called that, which Wells legitimized is bottomless. It seems fitting that Shakespeare's spirit world of *The Tempest* is the explicit source of the invisible forces and beings of *The Forbidden Planet* (1956), although the highly contrived reappearance of Robbie the Robot in *The Invisible Boy* the following year discarded any such pretentions to high culture. That movie presented a cold war fantasy about the military takeover of the world by a satellite (it was released in the year of Sputnik) controlled by an evil supercomputer, which is prevented by a young boy called Timmie whom the computer has made invisible so that he might play without being observed by his parents. (Really, don't ask.)

The Invisible Man himself 'returned' in 1940 in the imposing (if concealed) shape of Vincent Price, appearing here in his first horror movie as an escaped murderer who has been made invisible by the brother of Wells's Griffin. Price reprised the role for a cameo in *Abbott and Costello Meet Frankenstein* (1948), but the comedy duo made the

Playing for laughs: Abbott and Costello meet the Invisible Man.

invisible man's acquaintance more fully in, yes, *Abbott and Costello Meet the Invisible Man* (1951). Like ghosts and spirits, an invisible man might be homicidal and sexually predatory, but is also a perfect vehicle for slapstick.

As James Parriott, a scriptwriter for the 1975 US television series *The Invisible Man*, admitted, '[w]e realized that there was something very funny about invisible people'. This series, starring David McCallum as a decidedly sane and benign scientist Daniel Westin trapped in a state of permanent invisibility by an experiment with mysterious red rays that went awry, initially set out with serious action-drama intentions, in which Westin is persuaded to act as a secret agent for the government. But the scenarios veered, with a certain desperation, ever more towards the comic caper as the series progressed. Westin wore a prosthetic mask so that McCallum was conventionally visible on screen for much of the time (but ready to strip when the going got tough). It was a tall order to sustain the conceit that he was invisible beneath the makeup. '*The Invisible Man* was really a one-joke show,' admitted its producer Robert O'Neill. 'The minute you've taken the wrapping off his head, you've seen the joke.' And once Westin was invisible, there was the same problem that Whale had faced of convincing the audience he was still there. 'We'd have him brushing up against furniture and bumping into potted plants', said O'Neill. 'He ended up as the clumsiest guy in the world!'

Invisible in America

When the African American writer Ralph Ellison wrote his classic 1947 novel of racial discrimination and alienation in modern America, H. G. Wells's story gave him such a well-fitting metaphor that he more or less just appropriated Wells's title as his own: *Invisible Man*. Now regarded as a seminal exploration of social exclusion, Ellison's book describes the journey of a young black man from a promising position at an apparently progressive college in a southern state to an outcast and revolutionary in New York City. The unnamed protagonist's good intentions are slowly warped by the hypocrisy and prejudice of those he encounters, curdling into a frenzied and sometimes violent misanthropy. In the process, the man discovers that he is, to all intents and purposes, invisible in society. When, at the start of the book, we meet him in his 'invisible' state, he comes close to murdering a passer-by after bumping into him at night and being called an insulting (presumably racist) name. 'It occurred to me', he realizes when he is on the point of stabbing his victim, 'that the man had not *seen* me, actually; that he, as far as he knew, was in the midst of a walking nightmare! . . . Poor fool, poor blind fool, I thought with sincere compassion, mugged by an invisible man!'

Ellison's protagonist becomes invisible not merely because he belongs to an oppressed minority but because he refuses to be bound by the expectations afforded that group both by the powerful majority and by the minority's own members. Black people in America in the mid-twentieth century, Ellison says, are not inherently 'invisible'; but a college-educated black man who does not conform to the demands of white society (even – perhaps especially – the intellectual liberals and the supposedly 'colour-blind' revolutionaries of the New York movement called the Brotherhood) is too much the anomaly, too incomprehensible, too inconvenient, and must disappear. Like Dostoyevsky's narrator in *Notes from the Underground* (Ellison acknowledged the debt), he raves from within a hidden world, a recess that guarantees him meaningless omnipotence. 'I'm an invisible man and it placed me in a hole', he says – 'or showed me the hole I was in, if you will.'

Anyone familiar with the way invisibility is now used as a metaphor for the voices unheard in society may be surprised and shocked by the ferocity with which this image is employed by Ellison. His narrator is not some meek individual cast aside or trodden over by the strong, but is a man walking always on the verge of bloodshed. He belongs to no tribe, not even of the oppressed, but is an outsider. He is, in fact, a descendant of the magic-making juggler, the prestidigitator and trickster, the cunning thief and mountebank. And that is what American culture had forced African Americans to become in the mid-twentieth century, according to literary theorist Marcus Klein in 1964:

> Negro life contains the necessity for hiding, duplicity, treachery, for adopting shifting roles which the real reality goes on beneath. And the perfect metaphor for all the advice he has received is invisibility.

The invisibility of Ellison's narrator stems from the kind of unseeing later explored in Christopher Priest's *The Glamour* and China Miéville's *The City and the City*: a condition imposed by the selective vision of others, a manufactured blind spot painted over by the mind's eye:

> I am invisible, understand, simply because people refuse to see me. When they approach me they see only my surroundings, themselves, or figments of their imagination – indeed, everything and anything except me.

This predicament creeps up on you, the narrator says, and he explains the process using Wells's own faux-scientific metaphors:

> It came upon me slowly, like that strange disease that affects those black men whom you see turning slowly from black to albino, their pigment disappearing as under the radiation of some cruel, invisible ray. You go along for years knowing something is wrong, then suddenly you discover that you're as transparent as air. At first you tell yourself that it's all a dirty joke, or that it's due to the 'political situation'. But deep down you come to suspect that you're yourself to blame, and you stand naked and shivering before the millions of eyes who look through you unseeingly.

If one can scarcely read this passage today without thinking of Michael Jackson's cruel journey towards an 'unpigmented' existence before millions, no, billions of eyes – well, Ellison cast his net widely.

Ellison's Invisible Man explains that the condition of invisibility blurs moral boundaries: 'When one is invisible he finds such problems as good and evil, honesty and dishonesty, of such shifting shapes that he confuses one with the other, depending upon who happens to be looking through him at the time.' This is where Ellison catches the thread of fairy tale and fable that eluded Wells's cruder characterization: the invisible man lives *liminally*.

This marginal (in every sense), protean existence of the trickster figure is personified in Ellison's book by the character of Rinehart, an elusive figure for whom the narrator finds himself mistaken, although we never meet the real Rinehart and it's not even clear whether he truly exists. He is apparently a street shark, a pimp and bookie, a great lover, but also a church leader, at one point called a 'Spiritual Technologist' on the flyer for Rinehart's church that the narrator is handed:

> Behold the Seen Unseen
> Behold the Invisible
> Ye who are weary come home!

Perhaps this is what comes of that 'looking through' an invisible person: they become prismatic, many-faceted and many-hued, and spectral in every sense.

For all his dogged literalness, Wells wasn't blind to his own metaphor. He knew for himself, in a small way, what it was to be marginalized – he began as a young man from an ordinary home of modest means, trying to make his way in the refined world of letters. The young student in his semi-autobiographical novel *Tono-Bungay* (1908) explains that, on his arrival at college in London, '[i]n the first place I became invisible. If I idled for a day, no one except my fellow students . . . remarked it.' Joseph Conrad employed the same motif of invisible anonymity for his shabby anarchist Verloc in *The Secret Agent* (which Conrad dedicated to Wells), who disappears among the hordes on London's streets.

Invisibility is now the stock description for groups and behaviours that pass mostly unnoticed or ignored in society. As such, it implies an absence not just of visibility but of potency, voice, legitimacy. This usage risks diluting the force of Ellison's message, which is that invisibility is the price of nonconformity. You disappear when you make your existence impossible.

The invisible woman

The history of invisibility reveals that there are many ways to become unseen in society. Rather than arising from indifference, disdain or rebellion, invisibility can be imposed on one by fiat as a convenient convention. In the late eighteenth century, no one could seriously assert that women were unnoticed in European society; they were not, after all, even a minority. Yet they were (with the occasional exception of aristocrats) nowhere to be seen in positions of power and influence, whether in politics, the church, the arts or the sciences, because of course they were rigorously excluded from these places. Women were not supposed to aspire to such prominence, but were expected to blend into the background: to vanish unless needed. Invisibility was a characteristic of proper female decorum.

It was no coincidence that at the end of that century a popular feature of travelling shows of illusionism, juggling and magic was the so-called 'Invisible Woman'. Visitors would enter a room containing a long wooden box suspended on chains and evidently empty. At one end was a speaking trumpet through which a visitor could ask questions of the Invisible Woman supposedly lying in the box. He could place any object near to the mouthpiece and ask her to name it, whereupon the correct answer would come back immediately in a disembodied female voice. 'What astonishes and seems to derive from the marvellous', said a French periodical, 'is that nothing escapes the Invisible One. She sees and hears everything.'

How did it work? Speculation was rife, including the possibility that this was no illusion at all but involved genuine invisibility. But a pamphlet distributed in Paris in 1800 by a scientist named E. J. Ingannato

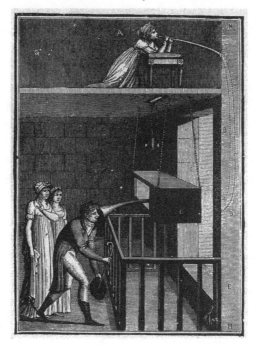

The Invisible Woman's secret unveiled.

revealed the secret. A woman was cosseted in the roof of the chamber, from where she could look down on the proceedings through a narrow slit. A speaking tube ran down into the wall of the lower chamber, from which the woman's voice could issue across a narrow gap to an amplifying horn in the freely suspended cabinet. Despite Ingannato's attempt at a spoiler, the show remained a sensation: at least three versions toured Europe in the early nineteenth century and Étienne Gaspard Robertson was still exhibiting the trick in 1830. 'To be stylish, one must speak in society of nothing [else]', proclaimed one magazine in 1806.

'The Invisible Woman Shows burst into the popular imagination at precisely the moment when women's public visibility was most fervently debated', says social historian Jann Matlock. One writer at that time said that fathers would give their fortunes to make their daughters as invisible as the lady in the box – after the Revolution, girls in France had the temerity to 'run around morning and night with complete liberty'. A female correspondent to a newspaper said that her husband was now demanding she should emulate the illusion and vanish. (All the same, she added, when invisible you can be ugly and still receive 'the most flattering compliments'.) Women were advised that invisibility was the key to sexual allure: as the French newspaper *Le Journal des Défenseurs de la Patrie* put it,

> This example should serve as a lesson to many women who are not
> invisible, and who are dropped as soon as they are known. Curiosity
> is the principle of love; it disappears as soon as it is satisfied.

Besides, as popular vaudeville songs remarked, invisibility could be a useful attribute for women planning to cheat on their husbands. Invisible women were socially acceptable, even alluring, but also untrustworthy.

Towards the end of the nineteenth century the Invisible Woman resurfaced in a new guise. Now it was not sufficient simply to express the ideal of a woman who lived her life unseen; the emerging public voice of women seemed to demand a more explicitly enacted banishment. She had to be *made* to disappear. In the 1880s the *Vanishing* Woman became a staple of stage magic. 'Never', says feminist critic Karen Beckman, 'had a magic trick been so prominent, so revered.'

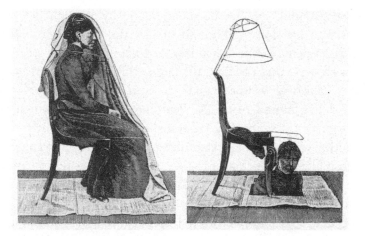

The Vanishing Lady, from Albert Hopkins's 1898 book of stage magic.

In a typical act, the woman would be seen seated in a chair, underneath which a newspaper was placed lest one suspect some hidden trapdoor. A drape was thrown over her, and then the magician, with a flourish, pulls off the covering – which until that moment had the form of the lady still visible beneath it – to reveal nothing but an empty chair. He might then call out into the theatre – 'Where are you?' – whereupon the woman rises from a seat in the audience to show herself.

Albert Hopkins tells how it was done. The concealed woman slides through the hinged seat of the chair and down through a trapdoor beneath – for the newspaper is a fake, with a flap that can be easily resealed with gummed paper. Meanwhile a wire frame shaped like the cranium holds the drapery in place, and is flipped behind the back of the chair by the action of pulling off the cloth. The woman hurries to some prearranged empty seat in the auditorium, slipping quietly into her place.

Beckman says that the success and popularity of this trick 'turns on the thrill of imagining momentarily that the female body can be disappeared without trace or consequence.' Even her reappearance in the audience, she says, plays into that image, serving 'to reinforce . . . anxiety about female power'.

The Vanishing Woman became so over-exposed that by 1896 John Nevil Maskelyne's deputy at the Egyptian Hall, Charles Bertram, was suggesting that it had lost its appeal. In that year Maskelyne sought to revive flagging interest in stage magic by introducing motion pictures into his shows, although in fact the movies probably just hastened its demise. But as we saw in Chapter 3, some stage magicians, such as Georges Méliès, turned to this new medium instead – and the Vanishing Woman made an immediate comeback on the screen. Méliès himself made *The Vanishing Lady* (1896), which simply shows the trick described by Hopkins, undermined by now-obvious stop-motion trickery. The British film-maker Robert W. Paul released an English film with that name in 1897, while Thomas Edison produced his own take on the theme in 1898. Arguably this implies that the anxieties about women in society had merely found a fresh outlet, but Beckman sees a more profound significance that is bound up with the very nature of cinema: 'The triple appearance of the vanishing lady in just three years suggests that she plays an essential role in the birth of film and that she embodies something about the medium itself.' Certainly the trope had staying power, most obviously in Alfred Hitchcock's *The Lady Vanishes* (1938), although Beckman also perceives the theme being played out in the 'fading star' movies of Bette Davis, *All About Eve* (1950) and *What Ever Happened to Baby Jane?* (1962). This obstinate female figure, Beckman argues, 'hovers endlessly between visible and invisible worlds'.

Dangers of the deep web

Aside from the politics of gender, one might see in all this a reflection of Griffin's discovery in *The Invisible Man* that invisibility is both a curse and a source of power. A part of the curse, in fact, is that the power is constrained by the very condition that creates it. This impasse is maddening. You can do what you want – but no one knows that it is you who has done it. The conflict drives Ellison's invisible man to insane excess, to pick fights for the slightest of reasons, enraged and exhilarated simultaneously that his opponent cannot even see the assailant.

This conundrum seems also to describe the predicament of the invisible internet troll, posting 'notes from the underground' that drip with venom and fury, his anonymity a passport to a power that can never be sated because it is illusory. What strikes one most about these characters, however, is that they are often not nearly as detached from human intercourse as the invisible men of Wells, Ellison and Dostoyevsky: the condition of detachment applies *only to their invisible selves*, the personas rendered distinct from their creators by a baroque pseudonym. It is true that these individuals commonly seethe with resentment and insecurity, but in normal life those characteristics are integrated with a person who holds down a job, talks to friends, perhaps raises a family. They are often repentant when tracked down and confronted, for (they think) the amorality of their actions belongs only to the persona of the invisible troll. As Plato predicted, invisibility has here eroded the moral norms that restrain the public self.

Linguist Claire Hardaker, a specialist on internet abuse, has explicitly drawn this connection between anonymity and what she calls the 'Gyges effect': 'the way that the internet can encourage a disinhibition people simply would not experience face to face'. But to explain how a faceless internet user turns into a malicious, invisible troll, she also invokes what one might call the Ellison effect: a sense of disenfranchisement brought about by social marginalization, economic disadvantage and lack of opportunity.

Digital gurus who sing the praises of cyberspace often show little interest in these matters, defending internet anonymity as an essential freedom. According to the American writer Cole Stryker,

> Computer technology has . . . empower[ed] individuals to redefine themselves in a social environment, to hack into their personhood, their identity, and truly become who they want to be. It doesn't matter if you're ugly or physically disabled – no one needs to know. And that freedom is contingent on the ability of Web users to take control of their identities – to be anonymous or pseudonymous [which is to say, invisible].

The callowness of this vision ('no one needs to know' that you're 'ugly') doubtless goes hand in hand with an inability to reflect on the unintended implications of jargon such as 'to hack into one's personhood' (with what kind of weapon?). For Stryker, the existence of internet trolls means you just have to toughen up. It's simply a corollary of the so-called GIFT (the Greater Internet Fuckwad Theory), which one source expresses thus: 'normal, well-adjusted people may display psychopathic or antisocial behaviors when given both anonymity and a captive audience on the Internet.' This 'theory' has plenty of empirical support, but it's not always funny. According to the managing editor of the Huffington Post, Jimmy Soni, thanks to anonymity 'comment sections can degenerate into some of the darkest places on the Internet'. As a result, the Huffington Post now demands that commentators post under their real names.

Randi Zuckerberg, a marketing director of Facebook, puts the problem a little more delicately than in the notion of GIFT:

> I think anonymity on the Internet has to go away. People behave a lot better when they have their real names down . . . I think people hide behind anonymity and they feel like they can say whatever they want behind closed doors.

But it's not really about names – anyone can invent a false one. It's about faces: about being seen. That has motivated the creation of Vine, a Twitter service delivered via video rather than message. It's one solution, perhaps. But the real question is not whether internet anonymity should be permitted, or whether it is some sort of fictive human right; it is whether we can understand how invisibility affects our behaviour.

Of course, the invisible underworld will always exist in cyberspace, as it does in the real world. There's a place some call the Deep Web, others the Darknet, the Undernet, or the Invisible Web. You need more than a search engine to reach it. And if you want to go there, you had better first don the invisibility cloak of some serious anonymizing software, because it's a dangerous, demonic world, full of all the things you probably – hopefully – don't wish to see, or to be seen seeing. Because sometimes we're right to shun and fear what, and who, lies unseen.

8 Vanishing Point

For the limits, to which our thoughts are confin'd, are small in respect of the vast extent of Nature it self; some parts of it are *too large* to be comprehended, and some *too little* to be perceived.

Robert Hooke
Micrographia (1665)

There may be more intelligence in invisible animals than in the larger ones.

Pierre Bayle
Dictionnaire historique et critique (1702)

At least one invisible world really exists, and it surrounds us. The Dutch cloth merchant Antoni van Leeuwenhoek was arguably the first person to infer its existence when, around the middle of the seventeenth century, he used his microscopes to look at drops of pond water. He found that the liquid was teeming with life. Here were innumerable little 'animalcules', too small for the unaided eye to register, going about their inscrutable business.

Invisibility comes in many forms, but smallness is the most concrete. Light ignores very tiny things as ocean waves ignore sand grains. During the seventeenth century, when the microscope was invented, the discovery of invisibly small objects posed a profound theological problem: if we humans were God's ultimate purpose, why would he create anything we couldn't discern?

The microworld was puzzling, but also wondrous and frightening.

There was nothing especially new or unusual about invisible worlds and creatures – belief in immaterial angels and demons was still widespread. But their purpose was well understood: they were engaged in the Manichean struggle for men's souls. If that left one uneasy about inhabiting a universe in which there was more than meets the eye, at least the moral agenda was clear.

But the divine role of Leeuwenhoek's animalcules and their ilk was opaque. They seemed to indulge in their cryptic, wriggly ways every-where one looked: in moisture, air, body fluids. Within human semen – Leeuwenhoek studied his own, transferred with jarring haste from the marital bed – there were tadpole-like animalcules writhing like eels. In 1687 the German mathematician Johann Sturm suggested that disease is caused by inhaling such invisible animals in the air. The Jesuit priest Athanasius Kircher proposed that the plague might be caused by the mi-croscopic 'seeds' of virulent worms that enter the body through nose and mouth – just a step away, it seemed, from a germ theory of conta-gion, although the impossibility of seeing bacteria and viruses with the microscopes of that time meant that it was not until the late nineteenth century, through the work of Louis Pasteur, Robert Koch and their con-temporaries, that this theory became firmly established.

The discovery of the invisible microworld, then, seemed to disclose deep mysteries – medical, philosophical, theological. We are still coming to terms with them today.

Fine-grained

The idea that matter might be composed of particles and processes too small to see has a long history: the whirling vortices of Descartes and the corpuscles of Boyle and Newton are descendants of the atoms of Democritus and his mentor Leucippus in the fifth century BC. This primitive atomic theory was perpetuated in the poetic masterpiece of Epicurean philosophy *De rerum natura* (*On the Nature of Things*) by the Roman writer Titus Lucretius Carus, rediscovery of which in the fifteenth century vitalized Renaissance humanism. But this fine-grained nature of matter came to seem like a genuine 'invisible world' only once the microscope enabled us, first, to appreciate the intricacy with which it was wrought, and second, to identify life amidst the grains.

When Galileo used one of the earliest microscopes to study insects, he was astonished and repelled by the sights, writing to his friend and sponsor Federico Cesi in 1624 that

> I have observed many tiny animals with great admiration, among which the flea is quite horrible, the mosquito and the moth very beautiful . . . In short, the greatness of nature, and the subtle and unspeakable care with which she works is a source of unending contemplation.

This wonder at the delicacy of nature's invisible texture was echoed by the English scientist Robert Hooke, whose 1665 book *Micrographia* put microscopy on the map. It was a project he took over from his friend Christopher Wren on behalf of the Royal Society and it suited the incurably curious (not to mention ambitious) Hooke perfectly. 'By the help of Microscopes', wrote Hooke, 'there is nothing so small as to escape our inquiry.'

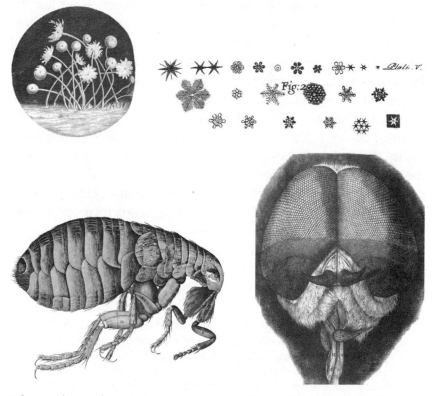

What Hooke saw through the microscope: mould, snowflakes, a flea and a fly's eyes.

Crucially, Hooke's volume was not merely descriptive. He included large, gorgeous engravings of what he saw through the lens, most of them skilfully prepared by his own hand (Wren may also have drawn some). The rhetorical power of the illustrations was impossible to resist. Here were fantastical gardens discovered in mould, snowflakes like fronds of living ice, and most shockingly, insects such as fleas clad in articulated armour like lobsters, and a fly that gazed into the lens with 14,000 little eyes, arranged in perfect order on two hemispheres.

This, Hooke concluded, was surely a demonstration of the infinite scope of God's creative power. In comparison, the finest contrivances of man – a needle's tip, a razor's edge, a printed full stop – looked crude and clumsy when seen up close. Invisibly small intricacy was considered the mark of divine artifice; as Hooke's colleague and sometime employer Robert Boyle wrote, it revealed God's providence 'in the hidden and innermost recesses' of things.

What most excited Hooke and his contemporaries was that the microscope seemed to offer the possibility of uncovering not just the invisible structures of nature but, in consequence, its concealed mechanisms. We saw earlier that, while in previous ages philosophers had attributed the causes of natural phenomena to the vague and insensible agency of occult forces and emanations, the mechanical philosophers of the seventeenth century argued that nature worked like a machine, operated by levers, hooks, mills, pins and other familiar devices too small to be seen. Now at last these structures might be disclosed. Henry Power, whose *Experimental Philosophy* advertised the virtues of the microscope a year before Hooke's magnum opus, wrote that we could expect to see at last 'the magnetical effluviums of the loadstone [magnet], the solary atoms of light, the springy particles of air'. Hooke insisted that 'Those effects of Bodies, which have been commonly attributed to *Qualities*, and those confesse'd to be *occult*, are performed by the small *Machines* of Nature.' Who knew what the microscope might ultimately reveal – perhaps even the traceries of thought itself?*

---

* In E. T. A. Hoffmann's novella *Master Flea* (1822), the hero Peregrine becomes entangled in a dispute between Antoni Leeuwenhoek and the naturalist and microscopist Jan Swammerdam over possession of a magical flea. This beast, in fact a quasi-human being, places in Peregrine's eye a microscopic lens that enables him to see through Swammerdam's eye the filigree of nerves that leads back into his brain, there ultimately to perceive his thoughts. As we saw in Chapter 3, the idea that

But Hooke and his contemporaries never found these invisible machines, and the resulting disappointment and disillusion contributed to the decline in microscopy in the early eighteenth century. The instrument showed an abundance of fine detail, but that detail couldn't be interpreted. As to why rhubarb purges but hemlock kills, the microscope was silent. Indeed, far from seeing the cause of qualities, microscopists found that those qualities themselves vanished at the smallest scales: opaque materials looked transparent, smooth objects looked rough and sharp ones blunt. Recognizing this, John Locke concluded that our faculties are not suited to comprehending the invisible world: God gave us only those powers needed to live in our own. 'For, to speak truly', he wrote in *Essay Concerning Human Understanding* (1690),

> yellowness is not actually in gold, but is a power in gold to produce that idea in us by our eyes, when placed in a due light . . . Had we senses acute enough to discern the minute particles of bodies, and the real constitution on which their sensible qualities depend, I doubt not but they would produce quite different ideas in us: and that which is now the yellow colour of gold, would then disappear, and instead of it we should see an admirable texture of parts, of a certain size and figure. This microscopes plainly discover to us; for what to our naked eyes produces a certain colour, is, by thus augmenting the acuteness of our senses, discovered to be quite a different thing.*

Likewise, Locke observed, coloured hair or silver sand becomes translucent in the microscope. And this is because '[t]he infinite wise Contriver of us, and all things about us, hath fitted our senses, faculties, and organs, to the conveniences of life, and the business we have to do here'. If this were not so – if one were to have the power of seeing directly into the invisible microworld (as some scientists in Locke's time believed Adam could do in Eden) – he would have too alien a concept of the world to be able to communicate it to us:

---

invisible thoughts can be visualized through technology has a long pedigree and is still alive and well.
* Many popular science illustrations could learn from Locke, showing as they do microscopic metal components of electronic circuits as though they were shiny lumps of solder. In fact the reflective lustre of metals disappears at small scales.

[I]f that most instructive of our senses, seeing, were in any man a thousand or a hundred thousand times more acute than it is by the best microscope, things several millions of times less than the smallest object of his sight now would then be visible to his naked eyes, and so he would come nearer to the discovery of the texture and motion of the minute parts of corporeal things; and in many of them, probably get ideas of their internal constitutions: but then he would be in a quite different world from other people: nothing would appear the same to him and others: the visible ideas of everything would be different. So that I doubt, whether he and the rest of men could discourse concerning the objects of sight, or have any communication about colours, their appearances being so wholly different.

This is a remarkable insight, for it dashed the prevailing assumption that the world of the very small is like ours but reduced in scale. Not at all, said Locke, it is utterly foreign and the human mind cannot be made to fit it. That injunction was often ignored subsequently, as we will see. But quantum physicists today, accustomed to the counter-intuitive rules of the atomic scale, would be unlikely to disagree.

A tiny hell

Hooke's *Micrographia* (1665) recorded life in the microscopic realm, but none that could not be discerned, with effort, by the eye alone: 'eels' in vinegar and mites in cheese. It was Leeuwenhoek's discoveries of invisibly small 'animals' – he was in fact seeing particularly large bacteria and single-celled organisms called protozoa – in 1676 (which Hooke verified a year later) that brought home with full force the teeming nature of the microworld. The anxieties about scales of perception that run through Swift's *Gulliver's Travels* make it clear how unsettling this was. In the land of the gigantic Brobdingnagians, Gulliver is disgusted by their bodies when seen so close up: 'Their skins appeared so coarse and uneven, so variously coloured when I saw them near, with a mole here and there as broad as a trencher, and hairs hanging from it thicker than pack-threads.' Living among the common folk of Brobdingnag he is repelled by the immense lice crawling on their clothes, possessing 'snouts with which they rooted like swine'. This is in striking contrast to Hooke, for whom the invisibly

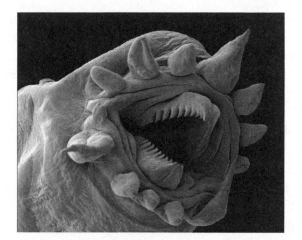

As this deep-sea hydrothermal worm indicates, the invisibility of the living micro-world can sometimes seem a blessing.

small becomes beautiful when rendered visible under the lens. The more common response to these gigantic fleas and mites was, akin to Gulliver, to find them grotesque: to experience a horror of the tiny world that modern microscopes seem to justify amply.

It was not just that these creatures looked bizarre and scary; the very invisibility of their pervasive presence was disturbing. This, rather than firm logical reasoning, is what tended to motivate a link between the animalcules and disease and decay. The sixteenth-century Italian physician Girolamo Fracastoro had already proposed that diseases might be transmitted by invisible corpuscles in the air, and even before Leeuwenhoek's discoveries some writers suggested that living agents of imperceptible smallness were the agents of the plague. In 1650 the physician August Hauptmann made the far-reaching claim that 'very minute and almost invisible creatures are the cause of all deaths in men and animals. The creatures are minute wormlets* beyond the reach of the unaided senses'.

Some of Leeuwenhoek's animalcules were parasitic worms and thus rightly suspected of being carriers of disease. But one shouldn't underestimate the influence of old ideas about the virulence of invisible demons on this quasi-germ theory. The narrator of Daniel Defoe's *Journal of the Plague Year* (1722) attests he has heard that if

---

* Worm could refer to any creeping or crawling creature; it is the etymological root of 'vermin'.

a person with the plague breathes on glass, 'there might living Creatures be seen by a Microscope of strange, monstrous and frightful Shapes, such as Dragons, Snakes, Serpents and Devils, horrible to behold'. He admits some doubts about whether this is true, but the message is clear: the invisible microworld is labelled 'Here be dragons'.

There is another factor at play in this reduction of the microworld to a familiar hellish bestiary. Lacking any prior notion of objects too small to see, the early microscopists had no option but to interpret these tiny forms in terms of macroscopic analogies, just as the mechanical philosophers expected to find hooks, levers and other everyday mechanics in the corpuscular microstructure that under-pinned occult forces. It was ever thus with scientific advance: the new and unfamiliar are interpreted by reference to the old and prosaic. It's no surprise, then, that microscopic observations tended to be depicted and interpreted through a haze of preconception, and occa-sionally outright invention, as the investigators forced their observa-tions into forms that they could understand. One can appreciate, for instance, why Hooke and others might describe the articulated cara-paces of mites and fleas in terms of 'armour plates', cuirasses and so forth. One can even find some sympathy with the claim of the Italian Giovanni Bonomo in 1687 that the mite resembles a tortoise. But in his book *An Account of the Breeding of Worms in Human Bodies* (1700), Nicolas Andry, a thoroughly competent microscopist and pioneer of parasitology, is obviously not describing exactly what he saw:

> The worms that breed in Humane Bodies, whether within or without the Guts, do oftentimes assume monstrous Figures as they grow old; some take up the Shape of Frogs, others of Scorpions, and others of Lizards. Some shoot forth Horns, others acquire a forked Tail; some assume Bills, like Fowls, others are covered with Hair, or become all over rough; and others again are covered with Scales and resemble Serpents.

There is just enough validity in these associations of the large and small for Andry's imagination to fill in the rest. And the point is surely that the creaturely analogies here are lowly, often noxious and loathsome: these little beasts are made so by the choice of comparison. Unlike Hooke, Andry evidently has no expectation that small is beautiful.

The link between invisible 'animals' and disease was very hard to prove – partly because standards of hygiene were so poor that threats of infection were everywhere, but also because many of the micro-organisms responsible for illnesses remained beyond the resolving power of the microscopes of the age. Nonetheless, the very possibility of using microscopes to see what was previously invisible offered assurance that afflictions like the plague were a part of nature, not the will of God. Whereas in 1625 the poet and writer George Wither could assert that the plague was a capricious agency that human intellect could never comprehend, by the latter half of the century it was widely assumed to be some form of 'effluvium': unseen and insidious, but no longer a moral issue.

The molecular demons

Littleness has been a consistent theme in the folklore of demons and fairies, where it enables puzzling and disturbing events to be ascribed to actions outside the bounds of human sight. But this kind of subvisible prestidigitation was by no means banished by the emergence of modern science. James Clerk Maxwell's demons are a testament to that.

The second law of thermodynamics, formulated in 1850 (see page 143), seemed to insist that the universe was moving steadily towards an accumulation of entropy and a spreading of disorder, culminating in a state of uniform heat wherein nothing could change for the rest of eternity. As a devout Christian, Maxwell could not accept that God would let this happen. Not only did it compromise eternal life, but the inexorable second law seemed to undermine free will. How could free will be rescued without violating thermodynamics?

Lucretius, while knowing nothing of entropy and thermodynamics, recognized that his atomism posed a problem of the same order: where, in the movements of these invisibly small particles along pre-ordained paths, might free will enter? In *De rerum natura* he solved that problem by animating atoms, permitting them to execute a random and unpredictable 'swerve' in their trajectories. Maxwell could scarcely attribute such semi-autonomous agency to molecules – but nonetheless his solution to the tyranny of the second law called for a similarly microscopic form of self-determination. His seminal work on the theory of gases convinced him that the second law is simply statistical. Gases contain molecules with a bell-shaped statistical distribution of speeds, the faster ones being in a sense 'hotter'. Temperature gradients are washed away because it is far more likely that the faster molecules will mingle with the slower, rather than by chance congregating into a 'hot' patch. There's nothing in the laws of mechanics to forbid the latter; it's just very unlikely.

But what if we could *arrange* for that to happen? Then the second law would be undone. We can't manage it in practice, Maxwell knew, because we can't possibly find out about the velocities of all the individual molecules. But what if there were, as Maxwell put it, a 'finite being', small enough to 'see' each molecule and able to keep track of it, who could open and shut a trapdoor in a wall dividing a gas-filled vessel? The 'demon' could let through fast-moving molecules in one direction so as to congregate heat in a single compartment, separating hot from cold and creating a temperature gradient that could be tapped to do work.

Maxwell laid out this idea in December 1867 in response to a letter from Peter Guthrie Tait, who was drafting a book on the history of thermodynamics. Maxwell told Tait that his aim was explicitly to 'pick a hole' in the second law – to show that it was 'only a statistical certainty'. The thought experiment offered a loophole that might reinstate free will.

Modern physicists now generally believe that Maxwell's demon (or a tiny mechanical device operating in the same way) cannot after all evade the second law, since the demon has to dissipate heat, and thus to generate entropy in its surroundings, as part of the process of gathering and storing information about the speeds of molecules. It turns out that Maxwell's ingenious but inadequate escape clause goes to the heart of profound questions about the relationship between information, computation and energy. Recent experiments have even been devised in which the human observer intervenes, demon-like, in a microscopic process and the entropic cost of doing so can be tallied. Here the demon itself has been reduced to a mere mechanism, if not in fact a metaphor. It is forgotten that, in Maxwell's day, invisibly small sentient beings going about their micro-business seemed possible, even likely.

Maxwell didn't intend his creature to be called a demon. That label was applied by William Thomson in an 1874 paper in *Nature*, where he defined it as 'an intelligent being endowed with free will, and fine enough tactile and perceptive organization to give him the faculty of observing and influencing individual molecules of matter'. Whether Thomson meant it or not, this seemingly trivial change of nomenclature connected Maxwell's being to a long genealogy of tiny or invisible spirits acting as agents and familiars with special powers, dating back to the *daimon* that allegedly advised Socrates. The devout Maxwell was not pleased with Thomson's terminology.

Now acknowledged as illusory, Maxwell's apparent victory over the second law might seem decidedly Pyrrhic in any case since, as Maxwell himself admitted, we couldn't possibly do what the demon does to separate gas molecules, one by one, according to their speed. Maxwell presumably could have argued that we might one day have the technological means to do so, but he didn't seem to hold out much prospect of that. There is, however, another way that his thought experiment could work: the demons might be real. Maxwell must have at least entertained this idea in order to take seriously the possibility that he had rescued free will. He never stated whether he regarded the 'demon' as a sentient being – 'Call him no more a demon but a valve,' he grumbled to Tait in response to Thomson's mischievous label, and he spoke also of a 'self-acting' device. But this was still a 'valve' or machine with thought and autonomy – or as Maxwell once put it 'a doorkeeper, very intelligent and exceedingly quick'. In any case, his touchiness about Thomson's quip seems rather puritanical even for a religious man until one realizes that Maxwell might have possessed a genuine belief in evil spirits.

Several of his contemporaries had little doubt that these 'demons' were to be taken literally. Thomson himself took pains to stress that the demon was plausible, calling it 'a being with no preternatural qualities [and which] differs from real living animals only in extreme smallness and agility'. William Crookes found another striking role for Maxwell's demon, arguing that it might in effect explain the mystery of radioactive uranium's seemingly inexhaustible supply of energy. At the annual meeting of the British Association in 1898 he suggested that uranium atoms might be like demons themselves, mining energy from the surrounding atmosphere by sifting hot gas molecules from cold. Although it isn't clear that Crookes thought any intelligent agency was involved here, he speculated elsewhere about sentient creatures inhabiting microscopic domains, their movements dominated by surface tension and the random buffeting of molecules.

Maxwell's subvisible demon shows how the projection of imagination into unseen realms still played a vital role in science. It's tempting to conclude that invoking invisible agencies like this amounts to no more than a disguise for ignorance. But Maxwell's thought experiment reveals how the impulse might serve precisely the opposite purpose: it clarifies and signposts where our ignorance resides, while at the same time bridging it, allowing science to press on despite inevitable gaps in knowledge. Invisible forces, agents and worlds in science show us what is missing, and what therefore needs to be added in order to close loopholes and broaden the reach of theories.

Small universe

Confronted with the microscopic subtleties of the natural world, Robert Hooke imagined this divine artistry continuing in infinite regression. 'In every *little particle* of matter', he wrote in *Micrographia*, 'we now behold almost as great a variety of Creatures, as we were able before to reckon up in the whole *Universe* it self.' Jonathan Swift, ever ready to poke fun at the speculations of the experimental philosophers, lampooned this idea in a piece of doggerel from 1733:

> So, naturalists observe, a flea
> Has smaller fleas that on him prey;
> And these have smaller still to bite 'em,
> And so proceed *ad infinitum*.*

* The flea, subject of one of Hooke's most celebrated illustrations in *Micrographia*, has long been the paradigmatic organism from the borderlands of the visible and subvisible. As such it often represents the point at which visible filth and vermin grade into invisible dirt, contagion and disease. Its liminal status is displayed more benignly in the flea circus, in which real fleas are traditionally harnessed like pack animals to operate tiny devices and machines: a carnival attraction that stemmed from the efforts of watchmakers to exhibit their skills at fine mechanical craftsmanship.

But there was no reason to doubt the possibility that Hooke raised, and which Blaise Pascal expressed more explicitly in his *Pensées* (1669):

Let him see therein an infinity of universes, each of which has its firmament, its planets, its earth, in the same proportion as in the visible world; in each earth, animals, and in the last mites, in which he will find again all that the first had, finding still in these others the same thing without end and without cessation. Let him lose himself in wonders as amazing in their littleness as the others in their vastness.

The discovery of subatomic structure and particles such as the electron, and of X-rays and other electromagnetic waves with wavelengths much tinier than those of light, reawakened this kind of speculation two hundred years later in a form that was almost unchanged. In *Two New Worlds* (1907) Edmund Fournier d'Albe envisaged an 'infra-world' at a scale below that which microscopes could register, peopled like Leeuwenhoek's drop of water with creatures ('infra-men') that 'eat, and fight, and love, and die, and whose span of life, to judge from their intense activity, is probably filled with as many events as our own'. In this world, Fournier d'Albe asserted, the electron is like an entire earth reduced in scale by a factor of ten billion trillion (10^{22}), while its own atoms are made of 'infra-electrons' that have their own 'infra-chemistry'. The human body, he estimated, could play host to around 10^{40} of these infra-men. Although this multitude is 'without the slightest net effect on our own consciousness', nevertheless the worlds are connected by the luminiferous ether, which can vibrate at frequencies relevant to both.

Fournier d'Albe was not blind to John Locke's caution against assuming that the invisibly small world was just like ours in miniature. Different forces would dominate, for example, and he suggested that electrostatic forces could constitute the gravity of the infra-world. But these were just substitutions – indeed, they were little more than analogies along the lines of the Neoplatonic correspondence of

---

Some later flea circuses substituted genuinely invisible motive forces, operating their carts and levers by means of electricity and magnetism.

macrocosm and microcosm. This flew in the face of all experience. The microscope had long shown that, even at a scale barely subvisible, the structures of the micro-universe are quite unlike those at everyday sizes: we can rarely understand their function transparently from their form. The universe does indeed have a hierarchy of scales, but these are not like Russian dolls, each a miniature replica of the last. As the British psychiatrist Henry Maudsley (who regarded occultism as a regressive pathology) noted in 1896,

> The universe, as it is within [man's] experience, may be unlike the universe as it is within other living experience, and no more like the universe outside his experience, which he cannot think, than the universe of a mite is like his universe.

It has been terribly hard, however, for even scientists to shake off these analogies of scale. Four years before Fournier d'Albe published his fanciful book, the Japanese physicist Hantaro Nagaoka had proposed a 'planetary' description of the atom in which electrons orbited a central nucleus like the rings of Saturn. And after Ernest Rutherford discovered in 1909 that atoms really do have extremely small, dense central nuclei, it became common to envisage atoms as

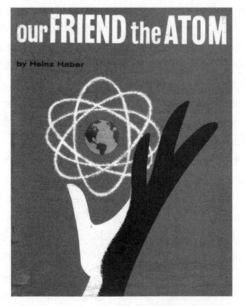

The atom as an invisible world, from a popular book of 1956.

akin to solar systems, with electrons orbiting like planets around the sun. This remains the iconic image of the atom, however inappropriate we now know it to be. It isn't hard to understand why Fournier d'Albe imagined that nascent atomic theory revealed the existence of worlds within worlds.

This habitual, anthropocentric macro-fallacy persists today. In what one hopes is a harmless pedagogical trope, it is common for atoms and molecules to be personified* – a gambit that can be traced back at least to Lucy Rider Meyer's 1887 book *Real Fairy Folks: Explorations in the World of Atoms*, in which the author, an American professor of chemistry, presented the chemical elements as 'fairy tribes' with physical attributes that reflected their chemical properties.

Meanwhile, the impulse to interpret nature's invisibly small mechanisms in terms of macroscopic machines remains strong among scientists. Molecules and nanotechnological machines are imagined as pistons, cogs and levers writ small – an imaginative leap that is not without some validity, but which may be actively misleading if taken too literally, for the molecular world is far stranger and more dynamic than we can easily visualize. When molecular scientists speak of molecular shuttles, motors, abacuses and computers, they are doing no more than what scientists have always done, which is to use metaphor and analogy to render the invisible in terms of the visible.

For example, Maxwell, William Thomson and their colleagues

Atoms have been presented as 'little people' from the Victorian age (left, showing "H_2O" in Lucy Rider Meyer's *Real Fairy Folks*) through to an image of the same molecule today (right). We no longer seriously consider the infra-world a place of sentient activity, but we can't avoid its legacy.

* I have often enough been asked whether electrons have consciousness to wonder what messages this sort of depiction really sends out.

concocted mechanical models of the ether that looked like little more than networks of bedsprings and piano wire, as though the God who conceived of the small and the immaterial were a divine Heath Robinson. Reviewing Oliver Lodge's *Modern Views of Electricity* (1889), the French physicist-philosopher Pierre Duhem found this decidedly British tendency all too much:

> Here is a book intended to expound the modern theories of elec-
> tricity . . . In it there are nothing but strings which move around
> pulleys, which roll around drums, which go through pearl beads, which
> carry weights; and tubes which pump water while others swell and
> contract; toothed wheels which are geared to one another and engage
> hooks. We thought we were entering the tranquil and neatly ordered
> abode of reason, but we find ourselves in a factory.

It is, of course, an invisible factory.

The seeds of illness

Once we see how early microscopists such as Johann Sturm, Nicolas Andry and Athanasius Kircher thought about disease, it might seem surprising that it took so long for the germ theory of infection and contagion to be formulated and that there was so much resistance to its acceptance. The story provides one of the best illustrations of how challenging it is for science to reach definitive explanations of the unseen – and why ultimately perhaps only seeing is believing.

Despite the fears raised about Leeuwenhoek's invisible animalcules, there was no compelling reason for him and his contemporaries to regard them as the cause rather than the effect of disease. Thomas Sydenham, perhaps the foremost authority on epidemics in seventeenth-century England, continued to hold to the miasma theory, in which tiny non-living particles in the air, perhaps emanating from the ground and from decaying and fetid substances in particular, were a kind of generic seed of illness. The idea prompted a valid belief in the importance of hygiene, but carried no notion of specific diseases having specific causes. It owed more to old notions of 'invisible forces' that influenced human health in vague, generalized ways. According to medical historian Roy Porter, 'Residual associations between contagion, astrology, magic and the occult explain the appeal of . . . the notion of "miasma".' What is more, these occult agencies continued to promote suspicions of diabolic and demonic intervention, and it is here that Porter locates the origins of the moral judgements about illness that are still evident today. 'Physicians' accounts of the causes of diseases have, in reality, doubled as sagas of condemnation', he says – in the same way, and for the same reasons, as accusations of 'magic' have done.

The miasma theory allowed no room for the possibility that diseases may be directly contagious, spreading from one person to another. That idea did not begin to gain traction until the end of the eighteenth century, for example in the proposal of the Scottish physician William Cullen that some sort of invisible 'effluvium' can be transmitted from infected individuals. Cullen also noted that contagion always spread the same disease – you didn't catch tuberculosis from a smallpox victim. Another Scot, Alexander Gordon, championed this contagion theory in the late eighteenth century, as did the American physician and poet Oliver Wendell Holmes decades later. But others resisted. Florence Nightingale's famous improvements in the military hospitals of the Crimea in the 1850s were predicated on a miasmic view in which the priority was simply to create a sanitary environment.

As Nightingale's case attests, we are so familiar now with the germ theory of disease that we can't help attributing to some of its pioneers insights and demonstrative powers that they didn't really possess. When John Snow traced an outbreak of cholera in London to a single water-pump in the Soho area and the handle was removed (albeit not by Snow) to prevent further infection, he might have established the foundations of epidemiology but he did next to nothing to clarify the cause of the disease. His miasmic critics could rightly point out that it was equally possible that piles of rotting refuse in the crowded area were responsible. Besides, Snow was unable conclusively to show that water from the pump contained dangerous microbes, and the intervention came only when the epidemic seemed to be declining anyway, as Snow himself admitted.

The possibility that tiny vermin had any causative role in disease was then still widely doubted. It was thought that such creatures were produced by decay, rather than having any active role in causing it. The Italian doctor Francesco Redi undermined the case for spontaneous generation of maggots in 1688 when he showed that none appeared in a piece of rotting meat covered by a gauze – the covering, he argued, prevented flies from laying the eggs from which the worms hatched. But microscopic organisms were much harder to physically exclude this way, so the question of their spontaneous generation remained unsettled. In 1748 the English biologist John Turberville Needham alleged that 'microscopical Animals' could be found in mutton gravy left to putrefy in a closed container, even after it was boiled. Even when the Italian Lazzaro Spallanzani showed that longer boiling times (required to kill the most robust of microbial spores) eliminated these organisms, he failed to end the debate. In any case, there was still no reason to think that such creatures played any part in the putrefaction itself and the ill health that could result from it. The eminent German chemist Justus von Liebig insisted in the 1830s that decay was a spontaneous, purely chemical process, and that microbes and moulds merely colonized the putrid matter opportunistically.

In another case awarded spurious retrospective clarity, the British surgeon Joseph Lister owed the success of his use of antiseptics more to luck than to genuine understanding. That doctors might reduce the incidence of infection during surgery by washing their hands had already been demonstrated in the 1840s by the Hungarian Ignaz Semmelweis, although his practices were rejected and ridiculed by his peers. In the 1860s Lister promoted a more effective antiseptic agent than Semmelweis's solutions of lime: carbolic acid or phenol, recently discovered as an extract of the coal tar produced by the gas lighting industry. Carbolic acid is a strong antibacterial agent and Lister said that germs could be killed off if doctors rinsed their hands and instruments with it, but especially if they sprayed the stuff (which is in fact rather corrosive and toxic) into the air. For Lister thought that disease was caused only by airborne germs, and he did not observe meticulous hygiene in all respects – he is said to have operated in an overcoat crusted with blood from previous procedures.

To Lister, then, the very air was hazardous. For hadn't John Tyndall, illuminating it with bright beams to make the invisible visible, shown that it was full of floating particles that might harbour germs? Of course, this could equally have supported miasmic ideas of non-living airborne particles, since it wasn't clear what these motes contained. Advocates of 'germ theory' like Lister could hardly make their case while the supposed agents of illness remained unseen. 'Where are these little beasts?', demanded the Edinburgh physician John Hughes Bennett. 'Show them to us and we shall believe in them. Has anyone seen them yet?'

Louis Pasteur and Robert Koch, the 'fathers of germ theory', were not the first to show them. For example, the Italian Filippo Pacini had already visually identified the cholera bacillus bacterium in 1854. The challenge was to show that such things were causes and not effects. Pasteur first did so with fermentation. As a professor of chemistry at Lille, he was aware of the immense importance of that process to the city's many brewers, vintners and vinegar-makers. It involved yeast, of course, but what exactly was this tan-coloured sludge? Using the microscope, Pasteur showed that, in contrast to what Liebig believed, it was alive. He also demonstrated that different varieties of yeast microbe could lead to different products of fermentation – some to alcohol, some to vinegar.

To challenge so grand a figure as Liebig, one might be thought to be either uncommonly brave or foolish, but Pasteur was neither. He was just exceedingly sure of himself. So deeply held were his convictions that he did not hesitate to report his findings somewhat selectively in favour of his hypotheses. His experiments allegedly ruling out spontaneous generation, for example, in which he showed that sugar solutions only underwent fermentation when they could be contaminated with microbes from the air, did not always work out as anticipated, since even Pasteur could not wholly avoid contamination when he wished to do so. But if a supposedly pristine solution began to ferment, then he recorded the experiment as 'unsuccessful'. This has been rightly exhibited as an example of poor science – for scientists should report all outcomes, regardless of whether their preconceptions are supported by them. But the truth is that science has often advanced this way, by extrapolation beyond what the evidence admits. Some scientists today might be yet more uncomfortable with the reasons for Pasteur's cherry-picking: he firmly believed that only God could create life.

Pasteur's revelation of the invisible agents of disease, however, came from studies not of human health but of the afflictions of the silkworm. In an example of the way social applications drive science forward (and not vice versa), Pasteur was called to the aid of the silk industry of southern France, which in the 1860s was facing ruin because of a disease called *pébrine* that was causing the worms to shrivel and die. The Frenchman Antoine Béchamp had spotted microbes on silkworm eggs under the microscope, and wondered if these might be lethal parasites. It can't be said that Pasteur's investigations were incisive – he was initially misled by a confusion between *pébrine* and another infection called *flacherie*, which is caused by a virus, still then an unknown class of micro-organism and invisibly small even to Pasteur's microscopes. But eventually he established the causative role of the parasitic microbe in *pébrine*, and most importantly, recommended effective hygienic measures for combating the infection: making silkworm nurseries better ventilated, for example, and regularly scrubbing their floors.

To establish germ theory, it was necessary to show that particular diseases were caused by specific, identifiable and unique microscopic organisms. This is what Pasteur and Koch achieved in the 1870s for anthrax. This deadly disease afflicted grazing animals such as cows and sheep, but could also be fatally transmitted to humans. It was a puzzle because it didn't seem always to involve contagion: animals could catch it without coming into contact with other infected beasts. Koch showed that the bacillus responsible for the disease could form long-lived spores that lie latent in grass until consumed. Pasteur completed the evidence in 1876 when he cultured anthrax bacilli in the laboratory and showed that they were deadly when administered to guinea pigs. He was assisted in this highly dangerous research by Emile Roux, who helped Pasteur to develop a vaccine from forms of the microbe weakened by heating.

From the 1880s, advances in the laboratory culturing of microorganisms by Koch, Pasteur and others led to the identification of the microbes responsible for many important diseases. In the early part of that decade Koch identified the bacteria that cause tuberculosis and cholera, while Pasteur found the invisible culprit for rabies and created a vaccine. Koch's assistant Georg Theodor Gaffky found the typhoid bacillus in 1884. It was by then irrefutable that diseases were caused by specific but hitherto invisible 'germs', and the consequences for their identification, treatment and elimination transformed health across the world.

★

Viruses, those other loathed pathogens of the microworld, are smaller than bacteria. They were discovered in 1892, when the Russian botanist Dmitri Ivanovsky showed that sap of a tobacco plant infected with tobacco mosaic disease remained infectious even when passed through the finest porcelain filters available. The sap, he concluded, must contain pathogenic organisms even smaller than bacteria, which are too large to pass through the filter's pores. Viruses remained invisible, inferred presences, however, until the invention of the electron microscope in the 1930s, which can 'see' more sharply than the light microscope. Investigations with the electron microscope showed that viruses are often exotic beyond the imaginings of conventional demonologies. No wiggling little serpents these; they exhibit a wide range of peculiar shapes, mocking our preconceptions about 'organic' form. Some are long cylindrical rods (such as the tobacco mosaic virus), others like Platonic polyhedral crystals or strange sea urchins, or equipped with arachnoid appendages and protrusions that enable them to navigate and inject their genetic core into infected cells. They are invisible demons appropriate for the sci-fi era: life (perhaps), but not as we had known it. Their lifestyles are equally unusual: they do not generally replicate by division, as bacteria do, but simply copy their genetic material and then assemble the protein coat that surrounds it. Some viruses produce their own enzymatic machinery for replication; others commandeer that of the cells they infect. Some, such as the AIDS virus HIV, encode their genes not in DNA but in strands of the related molecule RNA, the information in which is then written or 'reverse transcribed' into a corresponding stretch of DNA that gets incorporated into the genomes of the host organism.

This is why viruses are deemed to inhabit the very threshold of

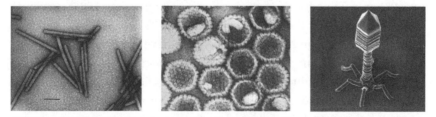

Viruses – invisible assailants until the 1930s – challenge the conventional morphologies of the monstrous microworld. From left to right: tobacco mosaic virus, Herpes simplex virus, and a virus that infects bacteria (a bacteriophage).

life. If life they are, it is life at its most stripped-down: replicating nucleic acids tightly packaged up in a protein coat, and barely anything else. Viruses are machines for copying themselves, achieving a deadly ability to respond quickly to their circumstances through the blindly refining sieve of natural selection. They show that if you can reproduce fast enough, you (at least, your descendants) can cope with almost anything. That is what makes many viral diseases, such as AIDS, so hard to combat. Throw at them whatever drugs you will, designed to disrupt some stage of the reproductive cycle – but the viruses will rapidly evolve an evasive strategy and you are back to square one. It is no more than natural selection at its most vibrant, but is hard not to see all this as the epitome of malevolence.

At war with the microworld

Regarding viruses as malevolent merely continues an old belief about invisibly small life. A century after the monstrous plague fantasies described by Defoe, but before the true perils of microbes were proven, the cartoonist for *Punch* magazine John Leech lampooned Leeuwenhoek in suggesting that London's thuggish underclass even infested its waters. The demonization of invisible beings became still more relentless after Pasteur and Koch revealed what we are up against. Efforts were made to communicate these hazards to the public, for example in theatrical demonstration lectures in which microscopic images of these devilish beasts would be projected onto a screen to make them the size of predatory dragons and serpents.

They became the new demons and ghosts, an omnipresent threat sanctioned by science. As the eponymous heroine's arrogant husband says to her in Theodor Fontane's realist novel *Effi Briest* (1895), scorning her conviction that their house is haunted, '[t]he fact that there are germs floating around in the air, as you'll have heard, is much worse and much more dangerous than all this spectral activity'. Children were (and are) taught that invisible 'germs' lie in wait everywhere, and they are enlisted, as imps and devils once were, to instil safe and sanitary behaviour. Now adapted to the fantasies of our age, viruses are 'alien invaders', and we go to 'war' on 'superbugs' with superpowers, repelling them like vampires with 'magic bullets'.

A DROP OF LONDON WATER.

a

b

c

d

Microbial monsters in popluar culture. *a*: The threatening microscopic 'pond life' of London, from *Punch* (11 May 1850); *b*: The invisible beasts are yet more monstrous by 1884, when this cartoon of a cholera microbe was published in the Parisian periodical *Le Grelot*. Here it is being unloaded from a ship in the port of Toulon, while a broker presents the bill of transportation, to be paid in the lives of the germ's victims; *c*: Poster for a theatrical 'voyage to the world of the infinitely small, seen through the giant electric microscope', from a Parisian theatre in 1883; *d*: Little has changed in this perception of the invisible microworld: here are 'germs' as depicted in current teaching resources by the Government of Western Australia Departments of Health and Education.

An educational text from 1912 promoting hygiene and cleanliness was called *The Human Body and Its Enemies*; the 'battle' against germs was one that people, especially housewives, were told they must fight daily. Like demons, these 'bugs' have fearsomely complicated names, and like demons they are thronging unseen and overwhelmingly all around us, just waiting until our defences are down. 'The microscope', warned the 1933 *American Red Cross Text Book on Home Hygiene and Care of the Sick*, 'has revealed the existence of innumerable little plants and organisms, so small that even millions crowded together are invisible to the naked eye.'

The discovery that not only the air but also the body teems with microbes introduced a new horror of intimacy into a society already made straight-laced by the strictures of Victorian etiquette. In the early twentieth century there were public-health warnings against kissing, touching, sharing clothes, even against handshakes. We were to keep our exudations to ourselves, for 'coughs and sneezes spread diseases'. Personal hygiene was a social obligation and bodily odours now carried a more shaming stigma. The antiseptic mouthwash Listerine, named after Joseph Lister but thankfully free of his favoured reagent phenol, was publicly marketed in the United States from 1915 (its genuinely antiseptic ingredients are various aromatic essential oils).

Microbiologist Abraham Baron put it plainly in the title of his 1959 book on microbiology: it is a case of *Man Against Germs*. When he explained that 'we share the world with an incredible vast host of invisible things', it was an admonition, not an expression of wonder. His description could have come straight from the pages of H. G. Wells:

> Their ways are strange, their life is alien, and their infinite populations overlap and encroach upon our human world. Germs are everywhere, wherever there are men, and they crowd densely in many places where men have never been. They float in the air, they rest in the soil, they swim in the saline waters of the seas and the sweet waters of lakes and rivers.

Even Baron seemed to understand that one need only replace 'germs' here with 'demons' to yield a statement our ancestors might have made. He begins his book by quoting the Talmud, which seems already to be equating devils with sickness:

The evil spirits crowd the public places, they crouch by the side of the bride. They hide in the crusts that are cast on the ground, and in the water that is partaken; they will be found in the oil, in the vessels, in the air, and in the diseases of men.

Nano nightmares

What has really brought all this teeming, invisible foe and filth into fine focus are the invisible rays: X-rays and cathode rays (which is to say, electron beams). Electron microscopes exploit the quantum wave-like behaviour of electrons to disclose form and shadow at scales where light can show nothing but a blur. The scattering of X-rays and electrons off crystalline substances, including viruses and biological molecules such as proteins and DNA, reveal the finest-grained structures meaningful to their biological activity: the positions of atoms themselves, a geography that Lucretius expected to be forever unseeable. And the invocation of invisible smallness as a source of wonder, delight and peril remains: a celebration in 2013 of the centenary of X-ray crystallography, invented by William Bragg and his son Lawrence, advertised itself as 'The Invisible Revealing of the Dangerously Beautiful'.

Fear of the potentially malevolent designs of imperceptibly small entities was evident in the initial public response to nanotechnology in the 1980s and 90s. Among scientists, this discipline is a loosely associated collection of attempts to visualize and manipulate matter on scales ranging from ångströms (the size of atoms) to hundreds of nanometres (the size of small bacteria). But in public discourse it became dominated in its early days by a single entity that nanotechnologists were allegedly aiming to construct: the nanoscale robot or 'nanobot'. This, it was said, would be an autonomous device patrolling the bloodstream for pathogenic invaders, or which would construct materials and molecules atom by atom. It was, in other words, a human avatar on an invisible scale.

What if nanobots ran amok, as robots are (in fiction) almost predestined to do? A rogue robot might be a menace, but it is at least a comprehensible one, a kind of superhuman being. A rogue nanobot, capable of replicating like bacteria and of pulling matter apart one atom after another, would be an unthinkable threat. Hidden from

An artist's fanciful impression of nanobots patrolling a human blood vessel. Notice that here the invisibly small craft are based on macroscale versions: the truth is that they haven't been scaled down at all, but rather, their surroundings have been scaled up. What we're really seeing is not an impression of the microworld, but a film set.

sight, it could reduce anything in seconds to a formless mass of atoms, which would then be reconstituted into an amorphous 'grey goo' of replica nanobots. The horror of this imagery, crudely but effectively exploited by Michael Crichton in his novel *Prey* (2002), was the same as that felt by medieval people surrounded by demons: how can you defend against a threat you cannot see?

The replicating nanobot was pure fiction. But scientists, distressed at how such misleading narratives can seize popular attention, need to recognize that they are not arbitrary, but rather spring from a deep cultural source, which is in this case linked to a history of aversion to the invisibly small. For if the grey goo images are frightful, they are also familiar. Invisible powers have long been held capable of animating shapeless clay, creating the fearsome Golem of Jewish legend, or of disintegrating and deliquescing matter and flesh, whether it is the Ebola virus or the mesmerism of Poe's 'The Facts in the Case of M. Valdemar'. What's more, the nanobot connects with long-standing visions of exploration of new worlds, most notably the voyage of the submarine *Nautilus* in which Captain Nemo explores the hidden deep sea in Jules Verne's *20,000 Leagues Under the Sea*. Once again, it seems we must remake the invisible microworld in our own image

Exploring the invisible microworld in *Fantastic Voyage* (1966).

before we can explore its promise and peril. This was clearly demonstrated by the 1966 movie *Fantastic Voyage*, based on a short story by Isaac Asimov, and the parodic 1987 remake *Inner Space*, in which humans are shrunk to a scale that allows them to navigate through the human body.

Letting out the genie

The extreme miniaturization of engineering that has its ultimate expression in nanotechnology has not yet given birth to an invisible nemesis and shows no sign of doing so. What it has done, in conjunction with the manipulation of invisible rays from Marconi's wireless onward, is to create an age of *technological invisibility*, in which things happen with no mechanism in sight, indeed even without our volition, embedded in an omnipresent field of information. Thanks to the shrinking of electronic and mechanical technologies, the occult cogs, levers and electrical switches of the machine – the same devices that once seemed to breathe life into ingenious automata in an indivisible blend of magic and mechanism – have become truly too small to see, allowing them to be packaged into smooth-contoured bodies that do wondrous things without a visible means of action. In this way, says Barbara Maria Stafford, we are drawn into inferring invisible operations and intentions from

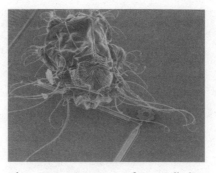

A bug wanders across the microscopic gears of a so-called microelectromechanical device carved into a silicon chip, as seen in the electron microscope.

visible effects, much as our ancestors did in the rituals of augury and thaumaturgy. With today's ultra-small technology, she says, 'interaction has once again become more disembodied and dependent on the actions of the invisible gods in the machine'. In previous times, this kind of invisible machination provoked suspicion of a perfidious ruse, a sleight-of-hand trick, or a hidden operator, perhaps a demon; but now, says Stafford, 'our anxieties have been flattened by the siren call of wireless products that make all operational functions compact, tinier, invisible'.

And so consumer items in stores now 'speak' to barriers, cash registers and computers in a language we can neither hear nor comprehend. Sensors control our cars and adjust our household environment, libraries leap magically into the device in our pockets. Dust, a metaphor for worthless matter while it was the smallest thing that could (just) be seen with the unaided eye, has become 'smart dust', a nano-technological promise (nothing more yet) of particles laced with invisible circuitry, programmed with the intelligence to self-assemble as we will them: to make a Golem, perhaps, rebranded now as a 'reconfigurable robot'. Take the Michigan Micro Mote, for instance: a complete wireless-enabled computer less than a cubic millimetre in volume – the size of a snowflake – which is capable of scavenging power from ambient sources such as light and heat in its environment. This kind of microprocessor would make air itself responsive, sensitive, smart. Or take the 'spray computers' imagined (but not yet built) by computer scientist Franco Zambonelli of the University of Modena and Reggio Emilia in Italy and his collaborators: as they put it, 'clouds

A computer the size of a sand grain: the Michigan Micro Mote.

of sub-millimeter-scale microcomputers, to be deployed in an environment or onto specific artifacts via a spraying or a painting process, [that] will spontaneously network with each other and will coordinate their actions to provide specific "smart" functionalities'. Such a spray, which could be practically invisible, would turn anything into a programmable, aware, quasi-sentient object. 'We could imagine', Zambonelli and colleagues say,

> a spray to transform our everyday desk into an active one, capable of recognizing the positions and characteristics of objects placed on it and letting them meaningfully interact.

The microscopic particles of such a spray, imbued with a capacity to sense, move and communicate, 'do not obey physical laws':

> Rather, they live in an environment for which programmers can be the creator of any desired virtual physical law. Thus, researchers are by no means limited to find[ing] out useful applications for existing self-organization phenomena but could, in principle, invent new mechanisms and laws.

New laws of a reconfigured nature. If that sounds like some powerful form of magic, it might come as no surprise to find that Zambonelli imagines these new mechanisms and laws to encompass what he calls a 'spray of invisibility', which we will encounter in the final chapter.

The final apotheosis of this (micro)machine power, says Stafford, will permit the ghost in the machine, at last, to step out of the box and wander the world. Diffused into everyday objects like clothing, jewelry, cash, paper, tables, chairs, and walls, the invisible 'smart dust' products of pervasive computing, we are told, will set up an intelligent force field. Yet when thus seamlessly embedded into mundane artefacts, will such technological mysteries be responsibly integrated into the spiritual, humanistic, and practical concerns of civil life?

Perhaps Stafford has begun to answer that herself. 'The typical modern "Enlightened" association of technology with secularization', she writes, 'tends to overlook its historical role in materializing the sacred.' This is true more than ever for the technologies of invisibility.

It has become a commonplace that advances like these would have seemed in earlier times to be truly magical. Less often acknowledged is how traditional reactions to invisibility can help us comprehend and negotiate the cultural changes that ensue. The boundaries between rationality and insanity can no longer be policed in behavioural terms. Is the person gesticulating and talking out loud in the street communing with demons of the mind, or with a friend via a hidden mouthpiece? Is the person fretting over the invisible threats of nearby radio masts succumbing to some modern version of the *mal aria* ('bad air') theory of contagion, or do they have a valid point? We entrust our digital secrets to the Cloud – a homonym for the mist often said to bring about magical invisibility – and we assume that this nebulous entity, not comprehended in the slightest by most users, can be summoned by technological wizardry to regurgitate them at will. We hold conversations with palm-sized boxes that divine our wishes. As we type on screen, invisible agents might read, or even edit, our words. With invisibly small technology we have tamed the ether – a necessary linguistic construct even if science has abandoned it – and thereby we have in a real sense animated the world.

9 Bedazzled and Confused

All moved under a cloak of invisibility . . . After you have seen this service uniform . . . you are convinced that for the German soldier it is his strongest weapon . . . Like a river of steel it flowed, gray and ghostlike.

Richard Harding Davis, war correspondent
News Chronicle, 23 August 1914

Visible form can only be distinguished when it is exhibited by differences of colour or tone, or of light and shade: with the reduction of such differences an animal or any other object becomes more and more difficult to recognize: in their absence it becomes unrecognizable.

Hugh Cott
Adaptive Coloration in Animals (1940)

Susumu Tachi, an electronic engineer at Keio University in Yokohama, has been making people invisible for several years. They vanish into the nondescript city landscape of Japan, urban ghosts taking on the appearance of doorways, passing trucks and pedestrians.

These disappearances are achieved with an elaborate paraphernalia of cameras, projectors and hooded reflective clothing so that the precise view from the subject's back can be captured and transmitted instantly onto his or her front – a technofix version (in the conventional pairing of science and fantasy) of Harry Potter's Disillusionment Charm, which makes the target take on the colour and texture of what is behind it. The key to the technology is a material that Tachi calls 'retro-reflectum', a carpet of little light-reflecting beads.

The 'invisibility cloaks' of Susumu Tachi.

What is an eye-catching optical illusion in Japan has become a political act in China, where artist Liu Bolin (dubbed the 'Invisible Man') paints himself (and occasionally other people) to vanish in front of government propaganda posters, images of police repression and modern consumerism. It is a comment on the social invisibility and marginalisation that Liu feels the regime engenders: in one work he 'vanished' six laid-off workers standing in front of the shop where they had worked all of their lives. But *every* culture, says Liu, shares questions about what and who is and is not seen, and he has also made himself invisible against scenes in the USA, France and Italy.

Chinese artist Liu Bolin uses paints applied with stunning precision to 'vanish' in front of a background.

Tachi, meanwhile, wants to extend his concept to architecture: to make a house with see-through walls. 'If we paint a wall [with the same retro-reflectum coating], then we can see behind it', he says.

An artist's impression of how Tower Infinity in South Korea will merge into the background.

'Even if there is no window in the room, we can see the scenery outside.' We are back with the optical trickery of Giambattista della Porta's camera obscura, projecting the world outside onto the wall within. The technology is different; the dream is the same.

Artifice of this sort – using an active light-emitting diode screen rather than a passive reflector of a projected image – is being planned to create the world's first 'invisible' skyscraper in South Korea. Called Tower Infinity and currently still on the drawing board, it has been designed by the US-based architectural firm GDS Architects. The building's construction has been approved, but there is as yet no scheduled completion date. The tower, almost 1500 feet tall, would be located in the Seoul suburb of Incheon. Cameras located at three levels all around the building would record the surroundings, and these images, computer-corrected to merge into a seamless whole, would be reproduced in banks of LEDs on the building's glass façade, so that it would act as a gigantic television screen that shows viewers what they would see if the building were not there.* As Tachi's 'invisibility cloaks' make clear, this is at best a compromised invisibility, which even in principle works 'perfectly' only for viewers standing at a single location. If Tower Infinity works at all, it should

---

* 'Invisible architecture' is a much used trope, although it can mean many things. For some architects, it means sculpting non-tangible sensations in space, such as sound and smell. The French artist Yves Klein took the idea more literally, envisaging invisible buildings with walls made of jets of air: an impediment to physical objects but not to light. Although he applied for a patent for an 'air roof', these plans came to little, which is a shame. Encased by invisible walls, would we alter our behaviour, or would we act as if these surfaces were still concealing structures, protecting us with their occult presence – just as we now talk on mobile phones as though still afforded the privacy of an invisible telephone kiosk?

be an impressive gimmick. Whether it will be invisible in any real sense is another matter.

One can achieve something like Tachi's (and Liu's) effect simply by standing in front of a screen as an image is projected onto it. Our eyes automatically 'read' the body onto which part of the image falls as near-transparent. Yet this is a cumbersome and imperfect kind of magic, scarcely portable and again conveying invisibility only if the viewer remains obediently in the right spot. In principle one could begin to compensate for that defect by adding more projectors; but if the cloaked object moves, the array of projectors must move too. It's certainly not the way to stroll about unseen.

We could get more sophisticated and place the 'projectors' on the cloak itself, making it a screen, just as is planned for the Tower Infinity. I consider in the next chapter the technological demands of producing something close to real invisibility through such a system. At any rate, these schemes offer a very different approach to invisibility from that imagined by H. G. Wells, which was based on optical transparency and the suppression of light reflection and refraction. Tachi's invisibility is that of the chameleon (or at least, the popular conception of it): a merging with the background. It is a kind of camouflage. As the pioneers of military camouflage appreciated, this is a capability best learnt from nature, which has developed an extraordinary battery of hiding tricks over the course of evolutionary history.

This makes the invisibility conferred by camouflage sound very different from that of magical tradition – there is no retreat into immateriality, no clever manipulation of light. Only the mind is deceived, and even then not by an inability to *see* so much as an inability to *distinguish*: this is hiding in plain sight. The difference is genuine, but it barely matters, for the science and natural history of camouflaged invisibility returns in the end to myth and magic after all.

Hidden nature

It is ironic that the chameleon has come to supply the most idiomatic example of natural camouflage, since its colour-changing skin is not a particularly effective mode of concealment, and might not be used primarily for that purpose anyway. There is now good reason to think that, for many chameleon species, most if not all of these

Now you see me. . .? The chameleon blends in but imperfectly, and often not at all.

transformational abilities are geared towards mating displays or signals of aggression and territorialism. Some desert chameleons turn black in the cool mornings to absorb the sun's heat, then light grey to reflect solar rays as the day becomes hotter. Most chameleons show little concern for matching their colour to the background.

Far more impressive is the colour-matching apparatus of the flatfish, the closest natural analogue of the computerized camera-and-display cloak. Light sensors on the lower part of the fish's body register the colour and brightness of the surface over which it sits and this information is reproduced by a battery of pigment cells on the upper body, sometimes with astonishing fidelity.

A flatfish, almost invisible against the sandy seabed.

Cephalopods – octopuses, squid and cuttlefish – have sophisticated colour-change mechanisms that are thought to be involved in both camouflage and communication for mating and displays of aggression. Some octopuses have coloured cells called chromatophores – yellow, red and black – along with white light-reflecting cells called leucophores, both equipped with a kind of mechanical shutter that displays or hides them. Certain species will not only match the tone or colour of the background but might also mimic its texture, for example producing bumps on the skin that resemble coral. Some squid have reflective skins that can produce bright, prismatic colours. These hues are not easily attained with a palette of natural pigments, and so nature uses a more versatile system for creating them that relies on microscopic light-reflecting physical structures in the skin. Such 'structural colours' are also responsible for the bright plumage of some birds and the dazzling markings of butterflies and iridescent insects. They arise when light is scattered from an array of tiny objects (typically rods or plates of dense organic material) regularly spaced at a separation of about the same size as the light's wavelength (hundreds of millionths of a millimetre). This scattering elicits a phenomenon called diffraction, in which reflected waves interfere with one another. Depending on the angle at which they are reflected, light rays of a particular wavelength (and thus colour) interfere constructively when they bounce off successive layers in the array, boosting that colour in the reflected light, while other wavelengths interfere destructively and the respective colours are eliminated. It is much the same process that generates the chromatic spectrum in light glancing off a tilted CD, where the light is diffracted by microscopic pits on the disc's surface.

In squid these structural colours are produced by cells called iridophores, which contain platelets of a protein called reflectin arranged into reflective stacks. Change the spacing of the layers and the colour changes. This change is brought about by neurotransmitter molecules, which alter the amount of electrical charge on the protein plates so that they can sit closer or are repelled further apart. It's an exquisite mechanism: a soft chemical machine for making rainbows.

The battery of biological light sensors, colour-changing cells and switching mechanisms needed for these acts of camouflage imposes

Stacked plates of the reflectin protein in iridophore cells create 'tunable' reflective colours in squid.

a high metabolic cost: it's a heavy evolutionary investment in hiding. Most disappearing acts in nature are more cheaply bought, placing trust in a single design that has evolved to blend with the organism's usual habitat. Some butterflies, moths and insects disguise themselves as dead leaves, bark, twigs and other foliage. Such mimetic camouflage represents one of the most exquisite adaptations to a creature's environment, and it played a key role in persuading both Charles Darwin and Alfred Russell Wallace of the power and capacity of natural selection. Of leaf butterflies of the genus *Kallima*, Darwin commented admiringly that it 'disappears like magic when it settles in a bush; for it hides its head and antennae between its closed wings which, in form, colour and veining, cannot be distinguished from a withered leaf together with the footstalk.'

When an animal's natural habitat has a predictable appearance, a single choice of camouflage might suffice – as here for a moth camouflaged against tree bark.

However, natural camouflage was also advanced as a counter-argument to Darwin's evolutionary theory, for the results may be so good that some biologists could not imagine why natural selection alone would have produced them. The palaeontologist Richard Swann Lull argued in 1917 that, for *Kallima* butterflies, 'a much less perfect imitation would be ample for all practical purposes and we cannot conceive of selection taking an adaptation past the point of efficiency'. Lull evidently felt his intuition was sufficient for judging what is 'ample' in nature.

Champions of Darwinian adaptation have sometimes been equally subjective and hasty with their judgements about animal coloration, imagining that concealment and invisibility must be the only conceivable function. Take the zebra's stripes. Rudyard Kipling offered a Lamarckian rather than Darwinian fable in his *Just So Stories* – the stripes are an acquired adaptation to the environment – but with the same favourable result:

> After another long time, what with standing half in the shade and half out of it, and what with the slippery-slidy shadows of the trees falling on them, the Giraffe grew blotchy, and the Zebra grew stripy, and the Eland and the Koodoo grew darker, with little wavy grey lines on their backs like bark on a tree trunk; and so, though you could hear them and smell them, you could very seldom see them, and then only when you knew precisely where to look.

That the stripes allow the zebra to hide was generally taken for granted. In his 1940 book *Adaptive Coloration in Animals* the British zoologist Hugh Cott quotes another specialist as saying that

> In the thin cover described he is the most invisible of animals. The stripes of white and black so confuse him with the cover that he is absolutely unseen at the most absurd ranges.

But the assumption that the stripes are for concealment becomes hard to sustain when the zebra is seen in its normal habitat of grassland, and it's not clear that the markings are about camouflage at all. They might instead be deterrents for biting flies, or an indicator of fitness, or a means of heat regulation – the theories are many and there is

The zebra might be hard to see among light-coloured vegetation (left), but put it in grassland and the stripes seem to have no camouflaging function at all (right).

no consensus. This is often the case with the scientific understanding of camouflage: what seems intuitively obvious at first glance often turns out to be not so at all on close inspection. This aspect of invisibility is subtler than it seems.

At the heart of the debate over the zebra's stripes is the question of what invisibility itself means. We've seen already that there are at least two ways to disappear: to become perfectly transparent, so that light may pass through you without being in the slightest part changed or deviated; or to blend into the background and become indistinguishable from your surroundings. In a scientific sense, the first is a matter of pure optics, the second – which includes the notion of camouflage – is perceptual. Near-transparency is rarely an option for vanishing in nature, although some sea creatures achieve it; imitating the backdrop is more common.

Yet these are not the only ways of passing unseen. One popular theory of the zebra's stripes is that they obscure not by concealment but by confusion. A division of the body into seemingly random patches of highly contrasting colours breaks up the outlines of the creature itself: a predator sees *something* but is unable to interpret it. Against some backgrounds, one colour might stand out much more than the other, creating puzzling shapes. Hugh Cott explained all of this in his 1940 book, which remains a classic exposition on camouflage:

By the contrast of some tones and the blending of others, certain portions of the object fade out completely while others stand out emphatically . . . the effect of a disruptive pattern is to break up what is really a continuous surface into what appears to be a number of discontinuous surfaces.

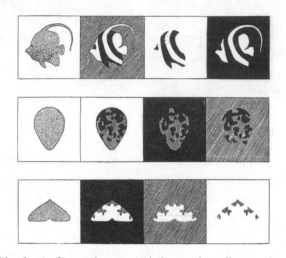

'Differential blending' of a patchy or mottled animal, as illustrated in Hugh Cott's *Adaptive Coloration in Animals*.

The stronger the contrast in tones, Cott explained, the more the outline of the creature dissolves into forms which 'tend to be interpreted as representing a number of different objects – none of which suggests, by its shape or arrangement, the body which bears them'. If the background is itself a patchwork of contrasting tones, then the mottled creature might itself blend in 'invisibly'.

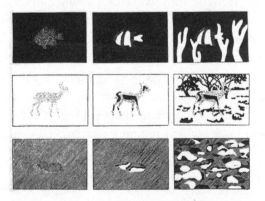

Cott argued that 'maximum disruptive contrast' hides the outline of the patterned animal. In a variegated environment (on the right) it can produce near-invisibility.

The art of hiding

The suggestion that some animals hide themselves via the 'disruptive' effects of their markings was first proposed in the late nineteenth century, but not by any scientist. It came from an eccentric American artist named Abbott Thayer, whose theories on coloration in nature became highly influential among zoologists.

Like many non-scientists who pronounce on science, Thayer felt that personal conviction amounts to a kind of evidence. He was highly strung, argumentative and staggeringly grandiose – it was said of him that he regarded 'God as a professional colleague (albeit a superior one)'. He risked undermining his perspicacious thinking about camouflage by turning it into a dogma that every fact had to be made to fit, whatever the contortions required.

Thayer's intuition was good, but in promoting it he was his own worst enemy. Not only did he tendentiously insist that only an artist, not a scientist, had the aesthetic sensitivity needed to understand biological markings, but he also insisted – in the face of clear evidence to the contrary – that *all* animal markings, even the most garish warning signals, served the purpose of concealment. 'All patterns and colors whatsoever of all animals that ever prey or are preyed upon are under certain normal circumstances obliterative', he wrote. 'Not one mimicry mark, nor one "warning color" or "banner mark" . . . nor any "sexually selective" color, exists anywhere in the world where there is not every reason to believe it is the very best conceivable device for the concealment of its wearer'. On occasion this invited ridicule, most notably for Thayer's claim that flamingos are bright pink so as to be camouflaged against the sunrise and sunset – as though invisibility at particular, brief times of the day were compensation for being strikingly visible for the remainder. His assertions of invisibility for some of the creatures photographed in his *Concealing Coloration in the Animal Kingdom* (1909) flew so much in the face of what the reader could plainly see that one can't help but think of the deluded Bulwer-Lytton parading 'unseen' before his guests.

Thayer lived in rural New England, where he immersed himself in the 'back to nature' ethos romanticized in the mid-nineteenth century by the writer Henry David Thoreau: sleeping outside all year round, studying birds and hunting for his supper. He was a talented but somewhat conservative artist, his pretty tousle-haired women all alle-gorical, virtuous and angelic, clothed in flowing robes and tunics, sometimes replete with feathered wings. His paintings speak of a soul ill at ease with the modern world.

But at the core of his unorthodox and sometimes bizarre claims about animal camouflage were some sound ideas. He was the first to suggest that hiding need not entail absolute invisibility: an animal could disappear by seeming to dissolve into random patches of colour that disrupt and hide the outline and so prevent identification. This became known as 'dazzle' camouflage – a rather inappropriate label, for it has nothing to do with the blinding effect of bright light, as invoked for the various 'invisibility' gems of medieval magic. 'It is based on the American slang "razzle-dazzle"', Hugh Cott remarked with disapproval, 'which has a meaning – expressive of active confu-sion – quite different from the correct English use for partial blinding by brilliant lights.'

Thayer also realized that animals could make themselves less visible with graded coloration that reduced natural brightness contrasts. A light-coloured underbelly and a dark sun-facing back reduce the effects of natural shadow and thus 'flatten' the body, making its outline harder to discern. This principle of 'counter-shading' became known as Thayer's law. Cott was convinced that it was the foremost secret of concealment. 'When at the same time the animal happens to be seen against surroundings with which it agrees in hue', he wrote, 'it will fade into ghostly elusiveness and become completely invisible from a short distance.' That is how a variety of fish, such as the barracuda and herring, hide themselves.

Thayer's ideas on camouflage were popularized among scientists by the English entomologist Edward Poulton, who described them in *Nature* in 1902. 'No discovery in the wide field of animal coloration', wrote Poulton, 'has been received with greater interest than Mr Abbott H. Thayer's demonstration . . . of the cryptic effect of the gradation of animal tints, from dark on the back to white on the belly.'

But Thayer's ideas were derided by Theodore Roosevelt, who became an enthusiastic big-game hunter after his US presidency. In 1911 Roosevelt published a detailed and merciless critique of Thayer, saying that 'the doctrine seems to me to be pushed to such a fantastic extreme and to include such wild absurdities as to call for the application of

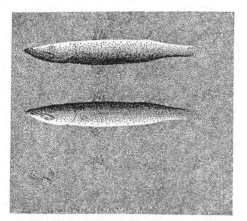

Counter-shading as illustrated by Hugh Cott. By offsetting a contrast due to light and shadow (top) with a contrast of pigmentation (middle), the animal can eliminate the tonal contrasts that pick it out from the background (bottom).

common sense thereto'. Roosevelt considered Thayer's insistence on the flamingo's 'concealing' pink shade to be the product of 'the obscure mental processes which are responsible for dreams'. On the African savannah, he said, 'no color scheme whatever is of much avail to animals when they move.'

Hidden forces

Despite such trenchant opposition, Thayer's reputation as an expert in camouflage led the US Navy to seek his advice after the outbreak of the Spanish-American conflict of 1898. This attention went to Thayer's head (as any recognition was apt to) and at the beginning of the First World War he petitioned his former nemesis's distant cousin, Franklin Roosevelt, to allow him to dictate the US Navy's policy on concealment. He had become convinced that at sea pure white was the most effective colour for making a ship invisible: 'only in the brightest moments is there any hope of distinguishing it from the sky' he wrote in 1916. This counter-intuitive idea was received with scepticism by naval officials, as were other variants on which Thayer insisted: a proposal to cover ships in billowing covers so that they might be mistaken for clouds – one can't help but think of Pooh Bear's attempt to steal honey from bees, published in the following decade – was rejected by Roosevelt as impractical.

★

As the epigraph to this chapter testifies, the First World War confirmed the view already promulgated in the Boer Wars that modern armies could operate more effectively when clothed in inconspicuous shades, rather than the aggressively strident colours of the past. The advent of aerial reconnaissance added to the argument: military convoys, personnel and artillery needed to be rendered inconspicuous from above. The French army was the first to develop these ideas, which is why camouflage is a French-derived word. The etymology is not quite clear, although in one plausible root is the Parisian slang term *camoufler*, meaning 'to disguise'. The French army formed its *section de camouflage* in 1915, headed by two artists: a portrait painter named Lucien-Victor Guirand de Scévola and the 63-year-old Impressionist Jean-Louis Forain. Its members were recruited from the theatre, sculpture and design: there was no question, for the French at least, that this sort of deception needed artistic sensibilities.

Thayer had those in abundance, but he could claim to be a specialist on natural camouflage too. And indeed camouflage came to be considered as much a science as an art, which even some artists acknowledged. The Royal Academician Solomon J. Solomon, a pupil of John Everett Millais, declared in *The Times* in January 1915 that '[t]o be invisible to the enemy is to be non-existent for him. Our attempts in this direction might well be a little more scientific.'

Solomon was engaged by the British Army as an adviser, and in 1916 he established a 'camouflage school' in Hyde Park, London. He advocated the use of netting to hide the outlines of artillery positions and designed observation posts disguised as trees with a steel core coated in bark. That trick was used so widely during the war that Charlie Chaplin used it for a skit in *Shoulder Arms* (1918), where he ran around in a tree costume knocking down German soldiers at the front.

Some scientists insisted that to make camouflage a precise and systematic business one should follow Thayer's lead and learn from nature. In Great Britain the Scottish zoologist John Graham Kerr promoted Thayer's theories of counter-shading and disruption – as Kerr put it, 'to destroy completely the continuity of outlines by splashes of white' – to the Royal Navy, with much the same obsessive vigour, and much the same lack of success, as the theory's author. Kerr sent the Navy countless memos and letters, but their experts felt that the lighting conditions at sea were too variable to make such proposals of anything more than academic interest. They settled instead for painting ships a monotone grey.

The Navy had a point. For it is the perpetual problem of camouflage that it tends to be context-specific: what works in one situation won't work in another. It's not hard to see how a scheme that will make a ship hard to see under the steely grey skies of the North Atlantic won't translate well to the azure Mediterranean. And how could one hope for an effect that conceals both in broad sunlight and at night?

That's why most military camouflage has tended to work by disruption rather than attempting perfect mimicry of the background. The British Navy finally came round to thinking that perhaps ships could be protected this way. But to Kerr's fury, they took this notion not from him but from the professional marine painter Norman Wilkinson, who conceived of the 'dazzle mode' independently and persuaded the Navy to adopt it. As a naval lieutenant in 1917, Wilkinson argued that efforts to make ships actually invisible were futile, and that instead the objective should be to confuse the predator – enemy submarines – rather than hide from them. 'The idea', he wrote that year,

> is not to render the ship in any degree invisible, as this is virtually impossible, but to largely distort the external shape by means of violent colour contrasts.

Wilkinson recommended to the officer in charge of the naval base at Devonport in southwest England, to 'paint a ship with large patches of strong colour in a carefully thought out pattern and colour scheme . . . which will so distort the form of the vessel that the chances of successful aim by attacking submarines will be greatly decreased'. Zebra-like light and dark patches would make it hard to estimate by eye a ship's position and course, and so frustrate attempts to judge the point at which its trajectory would intercept that of a torpedo. Wilkinson was placed in charge of a camouflage department, and his dazzle scheme was implemented on many naval ships and all merchant vessels longer than 150 feet.

The French cruiser *Gloire* painted with 'dazzle camouflage' during the First World War.

On a visit to the United States in early 1918 Wilkinson converted the American Navy to his scheme too. 'When I got over there', he wrote, 'I found they had five men farming invisibility schemes at royalties of about 100 dollars per foot run.' He got rid of all that nonsense, persuading the Americans to try out dazzle markings instead. In one test run Franklin Roosevelt, finding himself unable to guess a ship's direction, exploded 'How the hell do you expect me to estimate the course of a God-damn thing all painted up like that?' His admirals, fortunately, understood that this was the whole point.

After the war Kerr and Wilkinson became involved in a protracted dispute over who thought of the idea first. Kerr pointed out that he had proposed his system, based on sound biological principles, at the start of the war, but that the few instances in which it had been implemented (in the Dardanelles, for example, where Wilkinson had seen such ships in 1915) were bungled. Wilkinson denied any biological inspiration and argued that while previous schemes had aimed for invisibility, his sought confusion.

That was the real issue: was the purpose of the markings to hide the ship from sight or to make its movements incomprehensible? Kerr asserted that his aim all along had been disruption of perception, saying that this 'of course has a relation, but not a very close relation to invisibility'. Wilkinson, meanwhile, had generally stated from the outset that confusion was the objective, but betrayed a lingering confusion himself about whether this did or did not involve a kind of invisibility, saying that white paint should be used 'for those parts of the ship intended to be invisible'. The dispute was eventually settled in Wilkinson's favour, a judgement that Kerr contested (with some justification) for the rest of his life. Ironically, it seems that the ideas of both men had in any case been anticipated by accident, since ships had been painted in a patchy range of grey shades purely because of the vagaries of naval paint supplies. As the sea captain and novelist David Bone wrote in 1919,

> As needs must, we painted sections at a time – a patch here, a plate or two there – laid on in the way that real sailors would call 'inside out'! We sported suits of many colours, an infinite variety of shades. Quite suddenly we realized that grey, in such an ample range – red-greys, blue-greys, brown-greys, green-greys – intermixed on our hulls, gave an excellent low-visibility colour that blended into the misty northern landscape.

It was a further irony that dazzle camouflage might not have been terribly effective at protecting ships anyway. The statistics were equivocal and, as some naval commanders pointed out, conditions at sea are so changeable that what conceals in one context might reveal alarmingly in another. An inquiry in 1918 concluded that the camouflage made little difference to the attack rates on ships – but it did improve the morale of the sailors, happy in their belief that they could no longer be seen.

The same ideas, misapprehensions and ambiguities surrounded the concealment of land forces. During the First World War, armies still relied on the might of massed firepower, so invisibility was only valued for specialized assignments: snipers hid within hand-crafted camouflage, but the regular troops were offered no such benefits and had to trust simply to the reduced visibility of drab colours. Abbott Thayer recommended the use of camouflaged clothing aimed again at confusion, with loose flaps of cloth to break up the body's outline – a technique already used by deerstalkers in Scotland. It was rejected for general military use as being too complicated to manufacture.

That kind of disruption of the human form could also be found in the avant-garde art galleries of the day. The Cubism of Pablo Picasso and Georges Braque broke up the figurative into discrete patches of colour and shade, making a mockery of perspective and orientation.

It might have been pure coincidence that the dazzle-painted military vehicles and artillery of the First World War looked as though they had been conceived by the Cubists, but the analogy was not lost on those artists. The British Surrealist Roland Penrose recounted Picasso's response when walking one cold autumn evening with Gertrude Stein in the Boulevard Raspail of Paris:

> A convoy of heavy guns on their way to the front passed them, and they noticed to their astonishment that the guns had been painted with zig-zag patterns to disrupt their outlines. 'We invented that', exclaimed Picasso, surprised to see that his discoveries in the breaking up of forms should have been pressed so rapidly into military service.

'If they want to make an army invisible at a distance', Picasso said to Jean Cocteau, 'they have only to dress their men as harlequins.' Braque, too, was ready to claim credit: 'I was very happy', he wrote in 1949, 'when, in 1914, I realized that the Army had used the principles of my cubist paintings for camouflage.'

Military wizards

Making armies invisible through camouflage is an old dream. Julius Caesar mentions it in his history of the Gallic Wars, and it is by pre-empting Solomon J. Solomon's trick in making themselves indistinguishable from the trees of Birnam Wood that Malcolm's army advances on Macbeth at Dunsinane. The ability to vanish is prized in the martial arts: Japanese ninja assassins were renowned for passing unseen into the domains of their victims, and advanced masters of kung fu are said to be able to confuse their opponents so thoroughly that they cannot themselves be perceived. If your enemies are invisible (or merely thought to be), you may fall prey to paranoia, imagining invisible hordes on every hill. Solomon J. Solomon succumbed, becoming convinced that the Germans were hiding entire armies under vast nets. This has never been proved and seems unlikely, although some German sources after the war, perhaps wanting credit for such ingenuity, claimed that it was true.

As aerial reconnaissance and bombing acquired much greater importance in the Second World War, making things disappear became an urgent priority. In the early days of the war, any lessons learnt from the previous conflict seemed to be naïve and of doubtful value. The use of disruptive patterns in earth colours during the trench warfare had seemingly become ossified into the notion that invisibility was an innate property of the marking scheme – the standard-issue 'camouflage pattern' was thought to work as a kind of invisibility charm in any context, so that objects decorated in this way automatically vanished. In mid-1940 Hugh Cott, a disciple of Kerr, railed against absurdities such as the painting of the roofs of bright red buses in these patterns while they drove around Britain's cities. Such things, Cott said, 'bring down ridicule upon the art of scientific camouflage'. This talismanic attitude to invisibility was evident too in the way many soldiers, given netting with which to make an outline-hiding awning for trucks that should then be decorated with scrub and branches, merely cast it over the vehicles as though it were an invisibility cloak, ignoring the fact that they could still see the trucks perfectly well themselves. Netting adorned with green and brown patches designed for hiding in the countryside of Europe was parcelled out to soldiers in North Africa, who dutifully pegged these conspicuous cloaks around their vehicles in the yellow desert sand.

In explaining how to deploy these schemes properly, Cott added a warning, anticipating that some soldiers would be baffled when regarding the camouflaged objects at close quarters. Then they won't vanish at all, he said, but on the contrary will be 'glaringly conspicuous'. But 'they are not painted for deception at close range', he explained, 'but at ranges at which big guns and bombing raids are likely to be attempted.' This kind of invisibility is all a matter of perspective. When he was conscripted in the late 1930s to advise the Royal Air Force on camouflage, Cott found that in aerial warfare there was again a tendency to treat it as formulaic magic. Spitfire fighters had their upper surface daubed with patchy earth colours, allegedly making them invisible to attack from above, although it does no such thing for a fast-moving airplane at such height. Their underside was painted a light blue in the hope that it would merge with the sky.

In August 1940 Cott joined the Royal Engineers' Camouflage Development and Training Centre at Farnham Castle in Surrey. It was here that he discovered, one imagines not without alarm, the diversity of traditions in the art of making things vanish. For zoologists like

Cott joined a team that sought to accommodate not only the bohemian ways of modernist artists such as Roland Penrose* but also the
wiles of the stage magician Jasper Maskelyne, grandson of the famous
Victorian impresario John Nevil Maskelyne.

There is no better illustration of the continuity of the magical
associations of invisibility than the British army's appeal to Maskelyne.
He claimed that his illustrious family had a long tradition of 'helping'
the military: his grandfather John Nevil had, during the Boer War,
allegedly 'placed at the disposal of the War Office the results of
continued secret experiments with war balloons', while his father Nevil
had carried out experiments 'in connection with the flight of artillery
shells.' Jasper claimed that the Maskelynes had 'provided several magicians to assist Lawrence [of Arabia] with war-magic'.

It is hard to imagine anyone less suited than Maskelyne to a military
career, and the army may have come to rue their faith in the power
of invisibility magic once they discovered that their war wizard was
an incorrigible egotist and fantasist. At Farnham he was bored at
having to sit through talks on natural camouflage by the likes of Cott.
'For six weeks', he complained,

> I had to attend lectures where I learned how Arctic rabbits suffer a
> change of colour when snow falls, and why tigers hang about in tall
> grass. I had always believed that tigers hung about in tall grass for the
> same reason that boys hang about at street corners – on the chance
> of a pick-up.

'A lifetime of hiding things on the stage', he insisted, had taught him
more about invisibility 'than rabbits and tigers will ever know'. 'I
could, in fact,' he boasted, 'have hidden myself and most of the rest
of the class so efficiently that the lecturers would never have found
them in the duration of the war; but that would only have caused
trouble.'

After the war Maskelyne published an account of his exploits, *Magic:
Top Secret* (1949), which inflated his role shamelessly and made the

* Penrose published the *Home Guard Manual of Camouflage* in 1941, and gave talks
on the subject to military personnel which he enlivened with colour slides of his
glamorous wife Lee Miller lying on a lawn, her naked body concealed only with
netting, raffia and green paint.

most outrageous claims, including his having made entire cities vanish. By his account, the victory at El Alamein was due largely to his ability to make the British troops disappear and reappear at will. 'Magic helped to save Britain from German invasion in 1940', he asserted. 'For years, I and others made the mailed fist [of the Nazis] strike wildly against empty air, with considerable loss to its sense of balance.' Maskelyne proudly claimed to be on Hitler's 'Personal Black List', so that 'I took care to remain "invisible" as far as they [the Gestapo] were concerned.' He speaks of the larks of his 'Crazy Gang' of disciples in the European and North African theatres of combat in a way that makes it easy to imagine the generals wishing they could wring his neck.

Like the professional he was, Maskelyne gave away few of his secrets. Using 'some of the inner secrets of stage magic', he claimed to have hidden a machine-gun post in a field so that Lord Gort, Commander in Chief of the British Armies, couldn't see it even from a distance of six feet. He was a little more forthcoming about developing what one might call a form of early stealth coating (see page 251) that could hide aircraft from searchlights at night. By analogy with the black velvet backdrop used to hide objects on stage, Maskelyne painted the aircraft all over with varnish and, while it was still tacky, blew over it 'a special kind of black felt powder with a mixture of certain other substances used in stage work' – or more cheaply, with soot. 'Whether it would be possible', he said,

> by continued experiment, to produce substance [*sic*] that would give almost complete invisibility at a distance in strong light rays, I do not know, but I think it would.

He also mentions experiments with 'signalling to and from aircraft by means of "invisible" light rays', apparently meaning infrared radiation.

The successful camouflaging of British forces during the desert war owed less to Maskelyne than to the British filmmaker Geoffrey Barkas, who countered the aforementioned magical thinking about camouflage netting with a leaflet containing a comic poem about a hapless driver who employed his nets ungarnished with disruptive strips of cloth and hessian. Having refined his skills in the conflict in Northern

Ireland, Barkas was made 'Director of Camouflage' and posted to North Africa in 1941, with Cott as his instructor. For the Battle of El Alamein Barkas coordinated Operation Bertram, in which tanks were disguised as supply lorries while a fleet of dummy tanks misled Rommel about the direction of the Allied attack. If the construction of such 'fake scenery' sounds a likely job for a film-maker that was no coincidence, for like stage magicians they were seen as specialists in make-believe and illusion.

Steal away

Naval forces began the Second World War with no clear idea of what the previous war had to tell them. The efficacy of Wilkinson's dazzle approach had never quite been established and thinking was still divided between the principles of literal invisibility and confusion. The man who became regarded as the leading British authority on the subject – the Navy commander, artist and naturalist Peter Scott, son of the Antarctic explorer – never resolved that debate even in his own mind. In July 1940 he painted his own ship HMS *Broke* in blue, grey and white markings that he hoped would be disruptive by day while being invisible at night. Scott's design, called the Western Approaches Scheme with reference to its use in the North Atlantic theatre, was considered so successful that in May 1941 a memo within the British Navy called for other ships to be painted this way. The approach of using disruptive patterning in all pale colours seems, according to one naval report, to have indeed served a dual role: the light colours create low visibility at long range (as Thayer had argued), while at closer range, where the ship cannot hope to remain unseen (as Cott had explained), confusion is instead the result. Yet by 1944 the British had made invisibility the primary goal. The US navy, in contrast, adopted a variety of different camouflage schemes that sought primarily to disrupt visibility. Partly this reflects the differing backdrops of the British and American conflicts (the Atlantic/Mediterranean and the coastal Pacific) – but partly it reflects the irreconcilable ambiguities of when and how camouflage can work.

At the same time as engineers and designers were figuring out how to make ships, tanks and aircraft disappear, others were devising new ways of seeing them. For military forces in the modern age, invisibility is no longer just about evading light: their craft must also be hidden from radar, sonar and heat (infrared) sensors. Radar is a kind of seeing with radio waves, broadcast from an antenna and detected as they bounce back from an object. It was first discussed almost as soon as Heinrich Hertz discovered radio waves in the 1880s and found that they bounced off solid objects. The Serbian inventor Nikola Tesla recognized the possibilities for tracking large objects during the First World War, but these were only realized as an offshoot of attempts by the British Air Ministry to use radio waves as a 'death ray' that could shoot down aircraft. In 1935 this ambitious task was assigned to the meteorologist Robert Watson-Watt, a specialist on the radio detection of lightning. He soon concluded that radio waves could not be used for offensive combat (defusing fears that the Nazis had already developed such a weapon), but he proposed a better idea: using them to spot aircraft. Radar stations along the British coast provided advance warning of German aircraft raids that helped turn the tide in the Battle of Britain.

Hiding from radar is a matter of reducing the strength of the radio signal reflected back to the receiver, called the radar echo. In general the strength of the echo gives a rough indication of the size of the object causing it (although this also depends on, for example, what the object is made of and the angle of the incident radar beam). If an object can be made less reflective, it 'looks' smaller to a radar system. Stealth aircraft and ships designed to elude radar are typically coated with some substance – a paint, foam or fibre composite – that absorbs radio waves and converts the energy into heat. Radar is also strongly reflected by right-angle junctions of metal plates, which is why stealth craft tend to have a faceted appearance built up from sloping faces. This works fine for ships, but for aircraft it is not necessarily compatible with aerodynamic stability – the Lockheed F-117 Nighthawk stealth fighter, first flown in 1981, had a radar cross-section of about the same area as a drinks coaster, but was notoriously unstable in flight. Modern stealth fighters, such as the F-22 Raptor, have overcome this problem and to radar systems they look no bigger than golf balls or even large insects, making them impossible to distinguish from natural objects. Ensuring radar near-invisibility means accounting for all sources of reflection: radar will penetrate windows and bounce off surfaces such as instrument panels inside, unless the glass is made impermeable to radio waves by coating it with a transparent electrically conducting layer.

Invisible to radar: the Visby Corvette, a stealth frigate used by the Swedish navy (top), and the US Air force's F-22 Raptor fighter jet (bottom). Note the dazzle-type pattern on the frigate, albeit with relatively low tonal contrast.

The best form of optical camouflage is still debated even now. It's generally considered that plain colours, particularly grey, are the optimal compromise for ships in the open ocean, but in coastal regions dazzle-type patterning is still used for confusing the vessel's outline against a variegated background: the Swedish Visby Corvette frigate uses this for operation within fjords. Meanwhile, the British defence engineering company BAE Systems is developing a form of active camouflage using electronic ink, reminiscent of the colour-changing skin of the flatfish and the LEDs of Korea's Tower Infinity, that would adapt like a television screen to mimic the background. BAE has also reported a 'thermal invisibility shield' that will hide tanks from infrared heat-sensing cameras by matching the infrared appearance of the vehicle with that of the surroundings, using a 'skin' of hexagonal panels that can be heated and cooled very quickly.

Technology, then, has finally replaced the wartime wizards, offering a magic that really works. But magic is still how it is presented: 'It sounds more like a scene from a Harry Potter movie', says the sales pitch for the invisible tank, 'but BAE Systems is making the reality possible.' Because in the end, magic offers visions we can understand.

The ship that vanished

Perhaps the most notorious attempt to make a ship invisible during the Second World War was itself pure fantasy. The so-called Philadelphia Experiment is, like most conspiracy theories, not an arbitrary construct but a nexus of themes with deep roots: a potent little fable that concentrates associations and preconceptions around a shared cultural dream. Here we have the very real objective of military invisibility grafted onto the belief in an invisible order of reality mediated by electromagnetism, and coupled to the view of invisibility as a jealously guarded symbol of magical power.

The story is as painful a concoction of conspiracies as you will ever encounter, typified by the account of ufologist (and highly successful language teacher) Charles Berlitz and his co-author William Moore in *The Philadelphia Experiment* (1979). Through supposition, guesswork and the correspondence of eccentrics and cranks, Berlitz and Moore construct an ingenious house of cards. 'If in fact the Navy did somehow succeed, either by accident or design, in creating force-field invisibility', they ask, 'then what else might it explain?' The Bermuda Triangle (another of Berlitz's obsessions)? Space travel, secret mega-weapons? 'The possibilities are both endless and staggering!' Quite so.

For what it is worth, the tale generally runs like this. In the autumn of 1943, during a secret experiment called Project Rainbow, the US Navy used onboard electromagnetic equipment to make an entire destroyer, the USS *Eldridge*, vanish from a dock in Philadelphia. In one experimental test the ship was said to have been teleported to Norfolk, Virginia, and back again. Legend has it that Einstein's 'unified field theory' (never completed – or in the conspiracy theorist's version, suppressed because of its potentially horrifying applications) supplied the theoretical motivation for the experiment. (Feeble support is offered by the documented fact that Einstein did act as a consultant on explosives for the Navy during 1943–44.) Others say that a rival unified theory by Tesla, an Edisonian heir to the tradition of scientific stage magic, provided the impetus. Tesla died in 1943, surrounded by legends of secret 'death-ray' weapons that the military quickly confiscated.

The US Navy emphatically denies the story. According to the Office of Naval Research in Washington, DC, '[r]ecords in the Operational Archives Branch of the Naval Historical Center have been repeatedly searched, but no documents have been located which confirm the event, or any interest by the Navy in attempting such an achievement'. The ship's log from the summer until the end of 1943 records nothing but routine operations. 'There is no indication that Einstein was involved in research relevant to invisibility or to teleportation', the ONR notes primly. Well, they would say that, wouldn't they?

The Navy suggests that the legend might have arisen in part from the practice of 'degaussing', in which electrical cables were run around a ship to cancel the natural magnetism of the steel hull and thereby prevent it from triggering magnetic mines in the vicinity. These mines, deployed in the Second World War, contain magnetic sensors that detect a concentration in the Earth's own magnetic field induced by the presence of a large metal object such as a ship. 'It could be said that degaussing, correctly done, makes a ship "invisible" to the sensors of magnetic mines, but the ship remains visible to the human eye, radar, and underwater listening devices,' says the ONR.

We can now see the myth of the Philadelphia Experiment as a natural extension of the magical deployment of electromagnetic trickery in the nineteenth century for the purposes of stage-magical disappearances. It also creates a link between the old beliefs in invisibility magic and the modern dream that disappearances can be engineered by manipulation of the electromagnetic ether.

But *that*, at least, is no longer just a dream.

10 Unseen at Last?

Visible bodies may be made invisible, or covered, in the same
way as night covers a man and makes him invisible; or as he
would become invisible if he were put behind a wall; and as
Nature may render something visible or invisible by such means,
likewise a visible substance may be covered with an invisible
substance, and be made invisible by art.

> Paracelsus
> *Astronomia Magna* (1537–8)

Imagine there were no practical limits on the electromagnetic
properties of materials. What is possible? And what is not?

> Ulf Leonhardt and Thomas Philbin (2009)

In 2006, scientists working at Duke University in North Carolina
reported the first invisibility shield. It looked like this:

The astute reader will notice that this is very much work in progress. Even so, the device really is invisible – but only if you try to 'see' it by beaming microwaves across the plane in which it sits. That's to say, if you were to shine microwaves at the object from one side and detect them on the other, it would seem as if the waves have passed through nothing but air.

Now, is this invisibility in any meaningful sense?

That is how this new technology was reported in the media, with the endorsement and indeed encouragement of the scientists who devised it. The news stories were invariably embellished with accounts of Harry Potter's invisibility cloak, asserting that this was another example of science making magic come true.

This is an interesting image: magic, far from being the antithesis of science, is here paraded as its inspiration. This notion was explicitly voiced by John Pendry, a physicist at Imperial College and one of the key developers of the theory of 'invisibility shields', who has stated that invisibility was seen by him and his collaborators as a motivational 'grand challenge'. He was implicitly redefining invisibility, traditionally understood as 'hiding from sight', to mean instead 'concealment caused by redirection of electromagnetic waves'. But invisibility is not a technical term at all – like all magic, it has a symbolic and social function. When science borrows such terms, these associations will be imported too.

The curious thing is that sophisticated modern audiences were far more ready to accept these claims of invisibility than supposedly credulous medieval folk would have been. Our ancestors would have greeted the suggestion that this device *would* be invisible if one 'looks' in the right way with amusement and scorn. In fact, scientists themselves encountered similar scepticism at the outset in 2006. When optical physicists Vladimir Shalaev and Wenshan Cai first told their friends about the Duke University team's invisibility shield, they say that 'we were met with puzzled looks'. On reflection, they admit, 'the confusion was understandable'.

This is not to imply that such a disbelieving response is the more valid one, far less that the scientists were indulging in some kind of deception. Rather, it reinforces the suggestion of anthropologist Alfred Gell that magic, in its expression of hopes and desires, gives technologists something to aim for. This piece of invisible technology works in reverse, as it were: it demands that we accept on faith an 'invisible invisibility' that our eyes cannot verify. By dreaming magic into technological shapes, science fiction prepares us for that feat: the invisibility shields of the Starship Enterprise provide a datum on the graph to which this crude prototype can be connected.

We can see here too how, in modern science, the arcane has become unexceptional. In the nineteenth century much of the physicist's role was to make the invisible visible: to disclose what we cannot see via its effects on what we can. Now not even those visible effects are necessary. One can of course use instrumentation to demonstrate the passage of microwaves 'through' the invisibility shield, but it isn't something that can be experienced unmediated, in the way that one can hear a disembodied voice sent by radio transmission. The invisible can stay to all intents and purposes invisible, attested by nothing more than the authority of those who claim it. It is as though we have elected to believe the naked Emperor after all.

But he is not naked. It is more complicated than that, and we must attend carefully to what he is saying here – for however we choose to interpret it, it is surely something remarkable.

A new path for light

For you see, the field of science from which these invisibility shields have emerged, known as transformation optics, realizes an astonishing truth: that even seemingly fundamental principles of physics, such as the path of a light ray, can be reinvented and transfigured by the inventive design of materials. One can regard transformation optics, with only a little fancy, as the invention of an artificial ether for carrying light, governed by rules that the scientists themselves may prescribe.

The invisibility of H. G. Wells, in which light is not deviated by a substance because it has a refractive index equal to that of air, is possible in principle but not in practice, at least for an ordinary material. The invisibility promised by transformation optics takes a different approach: rather than trying to suppress the diversion of light rays, one must take command of that detour so as to restore the light to the path it would have followed if the object were not there. The rays, travelling from the source of illumination, are borne around the cloaked object and then brought together again on the far side so that, to the observer, it is as if they have travelled in a straight line all along. They pass around the cloaked object like a river flowing around a rock, reuniting smoothly downstream.

What manner of material will do this? None that is known in nature. Rather, the 'artificial ether' is called a metamaterial, made from 'atoms' consisting of tiny metal receivers and transmitters: coiled wires that pass the radiation to one another along a predefined route. The philosophy of metamaterials is to say that, if nature does not provide atoms that behave in the way we require, then we shall make our own.

This is a bold proposal, but it isn't quite how the challenge was first broached. In the 1960s, theoretical physicist Victor Veselago of the Lebedev Physical Institute in Moscow speculated about what might be required of a material for it to acquire a negative refractive index. We saw in Chapter 7 that the refractive index quantifies how much more slowly light travels through the material in question than through a vacuum. The vacuum (and roughly speaking, air) has a refractive index of 1; all ordinary materials have a refractive index greater than this, meaning that light is slowed down when it enters the material. It is this slowing that leads to the phenomenon of refraction, in which the light ray deviates from its original direction of travel.

But what can it mean for a refractive index to be *negative*? In physical terms, it means that a beam is bent the 'wrong' way. This would lead to a bizarre appearance for an object seen through such a medium: a pencil standing in a negative-refractive-index liquid, say, would seem not just to be kinked, but to be split cleanly in two. The pencil would look even odder if seen from above, seeming to float above the liquid surface.

Veselago explained what fundamental properties of the material were required to make these strange optics happen: one or other of the quantities called the permittivity and permeability would have to acquire a negative value. These quantities measure how strongly the material adjusts when it feels an electric (permittivity) or magnetic (permeability) field. If you place a substance – any substance – in an electric field, so that there is a positively charged electrode on one side and a negative electrode on the other, the charged particles it contains (the electrons and atomic nuclei) will tend to be pulled one

Ordinary refraction   Negative refraction

A material with a negative refractive index bends light the 'wrong' way (left), leading to strange visual illusions (right).

way or another: electrons towards the positive electrode, nuclei to the negative. Many molecules possess an intrinsic asymmetry of electrical charge, being more negative at one end than the other. In an electric field, the molecules will therefore feel a pull inducing them to line up in the direction of the field. (In materials that have no intrinsic charge asymmetry in the molecules such an imbalance can be induced by the electric field itself.) The permittivity is, roughly speaking, a measure of how easy it is for this sort of alignment to take place: the higher the permittivity, the more it is resisted.

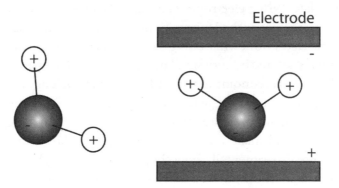

Many molecules, such as the water molecule shown here (left), have a so-called dipole moment: an imbalance of negative (-) and positive (+) charge on each side or end. This means that, if placed in an electric field, they will tend to be pulled into alignment with the field (right). The permittivity is a measure of how readily this happens.

The permeability is the equivalent quantity for a magnetic field: how readily the magnetic poles of the constituent particles of a material will become aligned when the material sits in a magnetic field. Because light is an electromagnetic wave – a wave of oscillating electrical and magnetic fields – the permittivity and permeability of a substance are an indication of how it reacts to the passage of light.

For either of these quantities to be negative sounds odd: it is as if an electric or magnetic field pulls the charges or magnets in the material the 'wrong' way. However, such things can happen naturally – some metals can display negative permittivity in response to short-wavelength light, for example. It turns out this means that they can essentially capture light on their surface, transforming it into waves called plasmons within the freely roaming electrons at the surface. But all transparent ordinary materials have positive values of both quantities. If either is negative, the medium is opaque.

Veselago wondered if it was possible to imagine a material that had *both* quantities negative. If that's so, then their product (permittivity times permeability) is positive, and this means that the material should still transmit light. In 1967 he published, in Russian, a paper describing what properties such a medium would have, and explained that these include a negative refractive index. His paper was translated into English the following year, but was regarded as a mere curiosity and forgotten for decades.

In the late 1990s, electronic engineer David Smith, then a postdoctoral researcher at the University of California at San Diego, came across Veselago's work while doing background research to understand how light is scattered by very small metal particles called nanoparticles. These behave as plasmonic materials for certain wavelengths of visible light: they have a negative permittivity, binding the light into surface plasmon waves. Smith figured that he could study this problem more easily if he could make a scaled-up analogue of the metal particles and measure experimentally how they responded to light. Since such a 'model metal' would have a much larger spacing between its 'atoms', Smith would also have to scale up the wavelength of the 'light' he shone onto it; by making the 'atoms' a few millimetres in diameter, he could use microwaves. That was an attractive prospect, because he'd studied microwaves for his PhD.

But what does it mean to make a material with 'atoms' as big as peas? Smith pondered that for some time, and eventually he discovered a paper published in 1996 by John Pendry and his colleagues containing just the prescription he had been looking for. Pendry presented a theory explaining that one can make an artificial plasmonic material for microwaves from an arrangement of little wires, which act like tiny antennas and receivers that absorb and radiate microwaves much as atoms do with visible light. It sounded ideal – except that the wires in Pendry's hypothetical structure were so thin that it would be almost impossible to handle them.

But as Smith pored over the paper, he realized that he could make structures with much the same properties as Pendry's wires from ordinary wire coiled into a series of loops, like a Slinky spring. With colleagues at San Diego he did just that, and found that it performed as predicted: like a plasmonic metal with 'atoms' a few millimetres in size, tailored to behave as no real atoms would. A name had already been coined for this kind of 'artificial material': it was a metamaterial.

Then Smith ran into Pendry at a scientific meeting and discovered that he was now imagining wire patterns that might produce more exotic behaviour – not simply the negative permittivity of a plasmonic metal, but a magnetic permeability that could also be tuned over a wide range of values, including negative ones. This was thought to be impossible, but Pendry argued that it could be engineered by using C-shaped wires called split rings. Smith was astonished. He realized that 'with the wires that could control permittivity and the rings that could control permeability, you could create materials with completely arbitrary electromagnetic properties' – a kind of tailor-made ether.

It turns out that the idea wasn't completely new. In the 1940s and 50s, researchers for the US Army had considered using similar arrangements of wires to build what was essentially a metamaterial, which they hoped might mimic how radio waves move through the ionosphere in the earth's upper atmosphere – coming back once more to Maxwell, Marconi and the magic of invisible communication. But that work never really progressed and was abandoned.

Now Smith, along with his colleague Willie Padilla, a postgraduate student at San Diego, started to wonder what else they could do with these loops and rings. What other electromagnetic responses could they build into their metamaterials? To explore that experimentally, they would need hundreds of the little 'meta-atoms' – too many to

A metamaterial for guiding microwaves, made from a grid of interlocking circuit boards imprinted with wires of copper foil.

make by hand. But Padilla found out that structures like split rings could easily be made on a printed circuit board by etching its copper foil into the required wire shapes. Indeed, their shape could then be controlled much more precisely than if they were built by hand, and the 'meta-atoms' could be easily assembled into an array by slotting the boards together.

The researchers figured out what wire shapes were needed to make a metamaterial that has, within a particular range of electromagnetic wavelengths, a negative permeability. This would make it opaque to microwaves of those wavelengths. They made it, and found the result they were expecting. But to prove that the permeability was truly negative, Smith and colleagues would have to measure the permittivity too. Otherwise they couldn't be sure that *this* wasn't in fact negative and causing the opacity instead. It was a tricky experiment, demanding specialized equipment that they didn't have. But Smith realized that, if he embedded the negative-permeability metamaterial in some medium that had negative permittivity, the opacity should go away and the stuff would be transparent. And he knew how to make a negative permittivity already: it was just his earlier plasmonic metamaterial. So the two types of artificial atom, made of loops and split rings, just needed to be interwoven like black and white squares on a chess board.

Again, the idea worked. But when Smith and Padilla submitted their results to a prestigious physics journal, it was rejected as insufficiently interesting. This prompted Smith to look carefully through the literature to see if anyone previously had thought about a material with negative permittivity *and* permeability. And he found Veselago's work.

Twisted light and warped space

Padilla ordered the paper from the library, and couldn't believe what he found. 'Every time I read through this paper I get more excited', Smith remembers him saying. It was at this point that the two researchers realized what they had made was a material with a negative refractive index. Making negative permittivities and permeabilities might be exciting for a small group of physicists, but a negative refractive index was truly weird and something comprehensible (after a fashion) to anyone who had ever contemplated the odd things that seem to happen to their anatomy in the swimming pool. But there was more, as Veselago understood. For example, the Doppler shift would be reversed: light coming towards you would have its wavelength stretched, while it would be compressed as the object recedes. It is as if the pitch of a siren were to fall and then rise, rather than the reverse, as an ambulance approaches and speeds past. For reasons connected to the reversal of the usual arrangement of electric and magnetic waves propagating through the materials, Veselago called them 'left-handed'.

The discovery of a negative refractive index gave the work a much more appealing spin, and when Smith and his colleagues resubmitted their paper in late 1999, it was accepted for publication. But to Smith's alarm his postdoctoral supervisor Sheldon Schultz (now a co-author of the paper) arranged a press conference before the publication, at the March meeting of the American Physical Society in Minneapolis. Smith was unsettled by this prospect of publicity. What if they were wrong? And in any case, he said, 'why would the general public be the least bit interested in something as arcane as a negative index?' In the end, he thought, all we're doing is scattering some microwaves off bits of metal – who would care about that? 'The night before the press conference', he says, 'I couldn't sleep, distraught and almost sick with the thought that we were making a big deal about our work when we might have missed something technical, or were overstating the significance.'

But he was wrong. 'As the time of the APS meeting and press conference drew near', he recalls, 'reporters started calling. And calling.

And calling. I had thought no one would be interested, but it seemed like everyone was interested.' They didn't necessarily understand the work – the *Washington Post* carried a meaningless headline about a left-handed material that 'reverses energy' – but they sensed it was strange stuff.

At that stage, even Smith and Pendry were still coming to terms with just *how* strange. The metamaterial was in effect directing light along paths that it was not 'supposed' to follow. To see why, we need to look again at the origin of refraction. Much confusion is caused by the statement that 'light travels in straight lines', because evidently it doesn't do that when it is refracted; it bends, or rather, it kinks as the beam goes from one medium to the other. But a more accurate statement is that light travels along the path that requires the *shortest travel time*. In empty space, that is indeed a straight line. But in a medium with a refractive index greater than 1, light is slowed down relative to its speed in a vacuum. So then the shortest path is one that reduces the amount of time spent in the slower medium. For a beam entering at an oblique angle, this means that the light can 'save time', relative to a straight-line trajectory, by travelling further through the 'fast' medium (such as air) and then turning to travel less far through the 'slow' medium (such as water). This is what causes refraction. The 'slower' the second medium (the higher its refractive index), the sharper the kink in the path, since then the time spent in the slower medium is less.

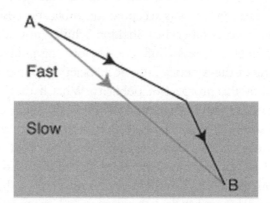

Light finds the path of least time. When it passes between substances of different refractive index, this means it takes a kinked route. By travelling a little further in the 'fast' medium and a little less far in the 'slow' medium, relative to the direct route (grey), the light beam can reduce its total time of travel from A to B.

This behaviour seems odd. How does the light beam 'know' which is the shortest path, before it has arrived at its destination? In other words, how does it know how much to bend? The answer is that light actually propagates over *all possible* paths between two points, but the waves interfere and cancel out almost completely *except for* those that follow the 'shortest-time' path.

Light waves can be made to follow non-straight paths even in empty space – if the *space itself is curved*. In this case, the light can't find a faster path than the curved one, because it is constrained by the shape of space, just as a car is constrained by a curving road and can't take an off-road short-cut without breaking the law (albeit the law of the road, not of physics). Curving space – or more properly, the fabric of space and time, called spacetime – is precisely what gravity does, according to Einstein's theory of general relativity. So light follows a curved trajectory when it passes close to a very massive object whose gravitational field distorts spacetime, such as a star. It was this prediction of Einstein's theory that was famously verified by astrophysicist Arthur Eddington during the 1919 solar eclipse, when he saw starlight curved around the occluded sun. Inside a black hole, meanwhile, spacetime is warped so much that light can never find a path leading out: all paths are closed.

In metamaterials, the ability to define the electromagnetic response more or less at will means that one can guide light along a trajectory that seems to defy the usual laws of physics. It doesn't truly do that – the path is still consistent with what the quantum theory of light would predict. Rather, it is the usual laws of *optics*, which emerge from the quantum picture, that are being broken here – and there is nothing in that deeper picture to prohibit that. But just as gravity involves a warping of spacetime that alters the path of light, so these metamaterials can be treated mathematically as though they involve a similar distortion of spacetime. Within the metamaterial, light is constrained to follow new and unusual routes in order to satisfy its principle of 'shortest travel time.'

It is worth pausing to take that in, because much else will follow from it. Optical metamaterials can be imagined as a reshaping of spacetime. They don't literally do that – not in the way a star or black hole does – but light behaves as if they do, mathematically and physically. It is as though the coordinates of spacetime have been redrawn, much like those now-familiar visualizations of general relativity which show spacetime as a rubber sheet marked with a grid and pulled into dimples by stars. It is because of this transformation of the coordinates of light's electromagnetic oscillations that this discipline, launched largely by Pendry and experimentally demonstrated first by Smith and colleagues, is called transformation optics.

Opening a hole

Bending light in unusual ways seemed like a useful trick of metamaterials. Smith and his colleagues started to think about making metamaterial structures that divert beams to different degrees at different points in space – by altering the shape and size of the split rings and loops, they could engineer a refractive index (whether positive or negative) that varies from place to place. That wasn't so hard: the researchers could just calculate what shape the rings and split rings had to be, and then use a computer-controlled process to etch the circuit boards. A slab of this material can act like a lens even if it is totally flat. Such 'flat lenses' can already be made from transparent materials with varying refractive index, and are used for example in photocopiers. But it's hard to vary an ordinary material's refractive index by very much; metamaterials would offer a wider range, and consequently greater lensing power.

Given this powerful possibility to tune the optical properties of the metamaterial continuously, everywhere, the researchers wondered what else might be made. This was when the invisibility cloak was born.

What Arthur Eddington observed in 1919 was a star *behind* the sun: its light was pulled into view because the sun's mass warped the surrounding spacetime. In other words, the starlight became visible not because it passed through the sun but because it curved around it.

Pendry realized that it should be possible to make a metamaterial structure that could guide *all* the light behind it in this manner: bending it around and reconstituting it at the other side. This amounts to opening up a kind of hole in space, where light cannot reach. According to Smith, it is

> rather like taking a pin and pushing it through the threads of a woven fabric, then moving it to make a hole. Light will move along the threads, going around the hole without any reflection. The light waves are bent

around the hole and restored on the other side. In effect, the space inside the hole – and any object within it – becomes invisible.

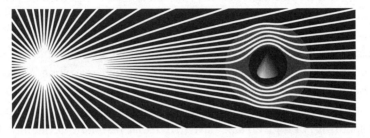

The diversion of light rays by an invisibility shield. The rays are guided around the cloaked object by the surrounding shield (grey) and then restored to their original path on the far side.

Pendry outlined his blueprint for a metamaterials invisibility shield at a scientific meeting in San Antonio in 2005. When Smith's students reported that talk back to him – Smith, now at Duke University in North Carolina, having been unable to attend – he realized that this was an application of his microwave metamaterials far more enticing than a negative refractive index. He proposed a collaboration with Pendry.

They weren't the only ones to have had the idea. When Ulf Leonhardt, a physicist at the University of St Andrews in Scotland, first heard about metamaterials and negative refraction in 2002, he too saw the possibilities. 'I thought, what could be the next big idea after the ideas of negative refraction are exhausted,' he says. 'It was immediately clear to me that this was invisibility.'

But it took Leonhardt three years to figure out how to make it work. In the summer of 2006, both he and Smith and Pendry, with Smith's colleague David Schurig, published papers back-to-back in the journal *Science* explaining the theory of how an invisibility cloak made of metamaterials might work. Their approaches were a little different, but as Leonhardt explains, 'the central idea of Pendry's and my version is the same: with optical materials one can implement coordinate transformations which can be used for "transforming things away" – that is, making them invisible.'

That's the theory – but could it be done? It took Smith and his colleagues only a few months to figure out the elements needed to make a circular shield that would bend microwaves travelling within the same plane: a two-dimensional microwave invisibility shield. To build the actual structure, with 10 concentric rings of metamaterial ranging from about 5 to 12 centimetres in diameter, they took a few shortcuts, and so the 'cloaking' wasn't perfect – it was a little reflective, and cast a slight shadow. But nonetheless any object inside the shield was more or less invisible to microwaves of about 3 cm wavelength. That's the device we saw at the start of this chapter, and inevitably it caused a sensation.

The invisibility manifests only at this particular wavelength because it depends on the light exciting a resonance in the split rings – like an acoustic tone making an undamped guitar string vibrate in sympathy. And split rings of a given size have particular resonant frequency. This is one reason why achieving invisibility across a wide range of wavelengths – the entire visible spectrum, say – is difficult: it's hard to excite such a broad resonance. The other limitation of this shield, at least from a public relations point of view, was that it only worked for microwaves. If people were going to be truly astounded by invisibility, they needed to *not see* what is allegedly hidden. The report of Smith's experiment in *National Geographic* magazine said it all: 'Researchers announced today that they've built the world's first invisibility cloak, although the fine print may disappoint science-fiction fans. The device works only in two dimensions and only on microwaves.'

To make a three-dimensional cloak is possible in theory, although fabricating it and putting all the parts together would be a big challenge. Making it work for visible light is still more demanding, because that would require all the 'meta-atoms' to be shrunk to microscopic scale to match the shorter wavelength of the light.

Instead of getting more complex, however, these invisibility cloaks have become steadily simpler. This is because researchers have recognized that still less perfect or more limited designs can lessen the difficulty of fabricating the structures while still affording some degree of cloaking. One important simplification came from Pendry and his postdoctoral student Jensen Li in 2008. They pointed out that, rather than hiding an object in free space, it could be more easily concealed under a 'carpet cloak' draped over a flat surface, such that light bouncing off the cloak seems to be coming from the underlying surface. There are no resonating elements in this structure, so it will work for a wider range of wavelengths, and the 'meta-atoms' can be made from appropriately shaped and spaced pieces of an ordinary material such as silica – making it more possible to create small structures that will operate at or near visible wavelengths.

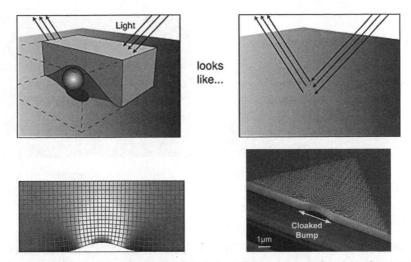

The carpet cloak of Li and Pendry. Here the refractive index of a slab of material concealing an object under a bump reflects light so that it appears to be coming from a perfectly flat surface (top). The refractive index of the slab varies from place to place (bottom left; shades indicate the size of the refractive index). A real version of this design was made in silicon in 2009 (bottom right).

In 2009 a two-dimensional cloak like this that worked at infrared wavelengths was made out of silicon, using the techniques employed to carve out integrated circuits, by Xiang Zhang at the University of California at Berkeley. The same fabric won't work for visible light, because silicon absorbs it.

Pendry and others have brought matters to an even simpler level. For a carpet-like structure in which an object sits on a surface and is hidden by a cavity, the hiding can be done with prisms and mirrors. The aim is to make a reflected ray look as though it has bounced off the underlying surface. One way to do that is to simply place another flat surface over the object. That alone won't fool us, because we can see that the new surface is nearer – that the space has a 'false bottom'. However, one can instead use two trapezoid-shaped prisms made from a transparent material to bend the ray in such a way that it looks as though it has come from the underlying plane. The trick is that the trapezoids must be made of a so-called optically anisotropic medium, which has different refractive indices in different directions in space. That's a property of some natural minerals, such as calcite: it is called birefringence and it gives rise to the double image of objects seen through them. So a crude invisibility carpet can be made simply from

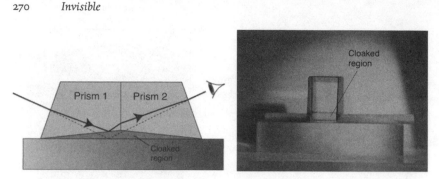

The calcite 'pseudo-cloak'. Using two prisms of a birefringent, transparent material such as calcite, light can be guided so that it seems to have been reflected from the surface underlying the hidden object (left). Such structures were demonstrated in 2010, here hiding a roll of paper (right).

blocks of calcite. It's not exactly true invisibility, because one can see the blocky, reflective 'cloak' quite clearly. But it is almost trivially easy to make once you understand the design, and it will work for all visible wavelengths of light. Two groups of researchers demonstrated this reduced form of cloaking at much the same time in 2010, one led by Shuang Zhang of the University of Birmingham in England and the other by Baile Zhang of the Massachusetts Institute of Technology.

You might think that this is starting to look like a Victorian stage trick, hiding an object behind smoke and mirrors. And although it can be couched within the mathematical language of transformation optics, in which the coordinates of the light's trajectory are shifted to produce concealing planes and surfaces, the connection to old-fashioned optical tricks becomes explicit when the idea is simplified even further. In the end, this sort of invisibility is a matter of guiding light around an object so that the background is visible in its place. Ordinary prisms will suffice for that. The result is a little crude, and again the prisms will glint and remain perfectly visible in air – but it is invisibility after a fashion, and at a scale limited only by the size of the prisms.

This was made clear in 2013 by Baile Zhang in collaboration with a team at Zhejiang University in Hangzhou, China, led by Hongsheng Chen. They created arrangements of massive prisms, made of optical glass, that would divert light around a central box-like compartment. Placed inside a fish tank, the prisms hid a fish inside the hole while pondweed could be seen swaying behind it. And a cat 'vanished' as it climbed into the cavity, while in the background a projected film of a butterfly flitting between flowers remained visible in the glass.

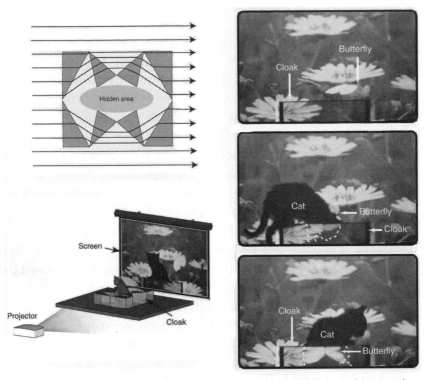

A cat (bottom) 'vanishes' into a cavity as prisms (top, left) direct light around it: transformation optics on the way to Victorian illusionism. . .

You can do the same thing with a series of mirrors, angled to direct into the foreground the reflection of the background. Now we have truly come full circle, and are back with the 'Living Half-Woman' of the music hall, truncated with artfully disported mirrors.

Carefully arranged mirrors were used to hide half of a woman's body in the Victorian 'Living Half-Woman' trick, as described in Albert Hopkins's *Magic* (1898).

World of Cloaks

This is perhaps an anticlimax to the high-tech visions of invisibility shields. But the illusions afforded, at least in principle, by transformation optics are many and dazzling and limited only by the imagination.

Just as every piece of magic should have its counter-magic, so it is possible to devise an 'anti-cloak' that cancels the effect of the cloak. A metamaterial anti-cloak might slot concentrically inside a cylindrical cloak, for example, to bring back into view the object that the cloak was hiding. This possibility has been demonstrated theoretically by Xudong Luo at Shanghai Jiao Tong Uinversity, Huanyang Chen of Suzhou University and their co-workers. One can also imagine building a metamaterial structure that, rather than sealing off a part of space from light, will open up a hole, in effect linking one region in space to another via something like the spacetime wormholes proposed by physicists as a means of time travel. The effect could be to conjure up in 'empty space' an image of a different location, such as the scene on the other side of a wall: a kind of invisible periscope.

A metamaterial cloak might also reveal *more* about the object inside: Pendry and his colleague S. Anantha Ramakrishna have shown that it can act as a magnifying lens. Pendry compares this with the way the light passing through a bottle of milk is scattered by the milk and the glass in such a way that the milk appears to go right to the edge of the bottle: you don't see the thickness of the glass itself. But a meta-material might make that image of the substance inside project, as it were, *beyond* the edge of the bottle and into empty space, so that it appears bigger than it really is. In 2008 Xudong Luo and his colleagues showed that, in theory, a metal surrounded by a metamaterial with a negative refractive index could act like this, becoming what they called a 'superscatterer' that blocks the passage of light over a wider area than its real dimensions should permit. In this way it would be possible to conceal an entrance in a wall, making it look as though the wall on each side extends across the opening. You would only be able to see the opening once you are close enough. Then we are back with Harry Potter and the hidden portal to Platform 9¾ at King's Cross Station: 'It is a bit like magic,' Pendry avers.

In principle, metamaterials can mould light almost as a potter models clay. As well as reconstructing light rays to look just as they did before they reached a cloaked object, these structures could manipulate the light that emerges to create the appearance of a *different* object – to make any arbitrary thing look like any other. This proposal for 'illusion optics' has been outlined by Che Ting Chan and his co-workers at the Hong Kong University of Science and Technology. Scale is no limitation – in principle a mouse could be disguised as an elephant, and vice versa. Calculating how the shape-changing metamaterial should be structured, let alone making it, would be a fearsome task, but there is nothing in principle to prevent it.

The analogy between transformation optics and the gravitational transformation of spacetime coordinates in general relativity suggests that exotic astrophysical effects might be mimicked in metamaterials. Huanyang Chen has shown how to use these artificial constructs to make light behave much as it would just outside a black hole's point of no return, called the event horizon. Xiang Zhang at Berkeley has suggested a way to build such structures out of real materials such as copper or semiconductors, while Ulf Leonhardt has outlined a scheme in which optical fibres rather than metamaterials may simulate some of the peculiar things that a black hole does to light.

Transformation optics is not just about light: at root, it is a science of waves. Depending on the scale of the cloak, one can in principle hide from light, microwaves and radar. But as Huanyang Chen has

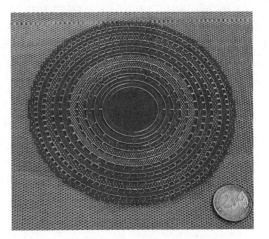

The acoustic invisibility cloak (with a one-euro coin to show the scale). The central cavity in this concentric structure is invisible to acoustic waves travelling laterally through the plastic sheet. (The holes in this structure have not yet been filled with the rubbery polymer needed to produce the full cloaking.)

shown, the idea can also be extended to sound waves: an acoustic cloak could make a submarine invisible to sonar. In 2012, researchers in Karlsruhe made an acoustic cloak of sorts from a metamaterial consisting of a sheet of PVC plastic drilled with a carefully configured pattern of holes and filled with a rubbery polymer. Bearing in mind that the acoustic waves here travel through the sheet as mechanical vibrations, it isn't hard to see how, by scaling up this structure, one can imagine making an invisibility shield for the seismic waves generated by earthquakes – hiding whatever is inside from their destructive effects.

Objects might also be hidden from water waves, allowing the waves to be guided around the object as if it were not there. Indeed, the entire flow of water – not just the surface waves – could be facilitated this way, so that the object leaves no disturbance in the fluid. A boat equipped with a shield like this could move through water without creating a wake. That would cut down on the drag that the boat experiences and would also make its passage much quieter.

Time bandits

The analogy between transformation optics, which redraws the grid for Maxwell's equations of electromagnetic waves, and the distortion of spacetime in general relativity permits a remarkable form of invisibility cloaking in which objects may be hidden not just in space but in time. In effect, this entails opening up a hole in spacetime – or to put it another way, editing out a bit of history.

The researchers who came up with the idea, physicists Martin McCall and Paul Kinsler at Imperial College, imagine a safe-cracker casting a spacetime cloak over the scene of the crime, so that he can open the safe and remove the contents hidden within the spacetime hole, while a security camera would record an empty room, without any 'jump cut' to betray the editing. Alternatively, by cloaking one's journey from A to B, one would appear to vanish from A and reappear instantly at B – as McCall and Kinsler say, creating 'the illusion of a Star Trek transporter'.

It should be clear by now that the imagery and metaphors that scientists choose to explain their work are not accidental. They tell us about the dreams that the science is cultivating. On the one hand, McCall and Kinsler are here evoking the futuristic magic of science fiction, in which mythical forms are dressed in clothes suited to the age of reason. On the other hand, the closest points of reference are still the ancient ones, in which invisibility serves as an agency of trickery, malevolence, thievery and deceit – in short, a challenge to morality.

How does one carve this hole in spacetime? The invisibility cloaks of transformation optics hide objects by bending light rays around them and then bringing the rays back onto their original trajectory on the far side. In contrast, the spacetime cloak would manipulate not the path of the rays but their speed. It would be made of materials that slow down light or speed it up. This means that some of the light that would have been scattered by the hidden event is ushered forward to pass before it happened, while the rest is held back until after the event. These slowed and accelerated rays are then rejoined seamlessly so that there seems to be no gap in spacetime.

The process can be illustrated by showing the trajectories of light in graphical terms, simplified to encompass just a single direction of space – light rays travelling in a line, as if down an optical fibre. If a ray travels along this space at constant speed, a graph of its position as a function of time is a straight line. If, however, the speed changes – for example, if it encounters a region of different refractive index – then the slope of the line changes, becoming flatter with speeding up and steeper with slowing down.

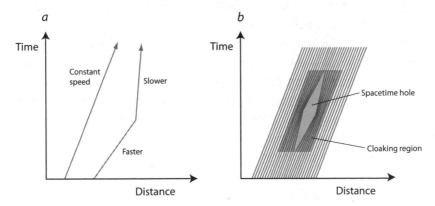

Spacetime plots of the trajectories of light rays. A constant slope corresponds to a ray moving at constant speed (*a*). By manipulating the speeds of the rays within a region of space and time, a hole can be opened up where no light penetrates (*b*).

Imagine now that a series of rays is moving sequentially through this space: there is, in other words, continual illumination. If, at a certain time, we begin manipulating the speed of the rays at different points in space, speeding some up, slowing others down, and then induce the reverse change of speed so as to restore the rays onto the spacetime trajectory on which they began, we can bend the trajectories to open up a hole into which no light rays can penetrate. Whatever goes on in that hole – which opens and then closes as time passes – does not affect the light streaming 'around' it. An observer outside this cloaking region – either further 'downstream' in space or later in time than the cloaking event, or both – sees nothing but the apparently unbroken, unperturbed succession of rays. However, the spacetime hole opened up by the cloak is not symmetrical: it exists in one direction but not in the opposite direction, even though the cloak itself could be made invisible from both directions. So an observer on one side might see an event that an observer on the other side will swear never took place.

To make a spacetime cloak, then, light must be slowed or speeded up relative to its speed before it enters the cloak. Now, if the region beyond the cloak is empty space – open air, say – then this is very demanding. We know that light is slowed if it enters a medium with a greater refractive index than air – in other words, pretty much any ordinary transparent material. But how can light move *faster* than it does in a vacuum? It turns out that this is in fact possible: in some exotic substances, such as ultracold gases of alkali metals, light has been speeded up by a factor of 300, so that, bizarrely, a pulse seems to exit the system before it has even arrived.

Such a peculiar and impractical medium is not, however, necessarily needed. If the cloak itself is surrounded by some transparent cladding material, then the light must be speeded up or retarded only relative to this. All the same, to obtain perfect and versatile cloaking one must alter both the electric and the magnetic fields of the electromagnetic wave. Most transparent materials, such as glass, are non-magnetic and so don't affect the magnetic field. Moreover, the effects on the electric and magnetic components must be the same, since otherwise some light will be reflected as it enters the material, making the cloak itself visible. When the electric and magnetic effects are equalized, the material is said to be 'impedance matched'.

There are no ordinary materials that would satisfy all these requirements, but metamaterials can. Because the cloaking now involves time as well as space, the cloak's optical properties need to change over the course of its operation, switching within each layer of material by the right amount at the right moment as the light passes through.

Just as birefringent minerals have been used to make approximate, imperfect but feasible invisibility cloaks, McCall and colleagues realized that sandwiches of ordinary materials with 'tunable' refractive indices might be used to make 'approximate' spacetime cloaks. For example, one could use optical fibres whose refractive index depends on the intensity of the light passing through them. A control beam would manipulate these properties, opening and closing a spacetime cloak for a second beam. The result is that although an object or event can be fully hidden, the cloak itself is not: light is still reflected from it. In 2011 a team at Cornell University worked out how to put this idea into practice, in effect enabling one light beam to hide from interacting with another during the fleeting 15 trillionths of a second over which a spacetime hole was opened up within an optical fibre.

This is, in truth, a long way from hiding a safe-cracking escapade. And although there are reasons why this kind of trickery could be useful for optical communications and computing, it is not going to enable a stage magician to teleport a rabbit.

The real thing

Might these sometimes glaring disparities between the apparent promise and the technological delivery of invisibility be telling us that we simply haven't pushed the science hard enough? That is what computer scientist Franco Zambonelli and his doctoral student Marco Mamei at the University of Modena and Reggio Emilia suspect. And so, like realists of the fantastic, they have pursued the legendary cloak of invisibility with dogged literalism to establish if these ideas really can supply a fabric that makes a person vanish.

It is hard enough, Zambonelli and Mamei say, to imagine making a solid wall from metamaterials that is truly invisible. How much more so, then, for a cloak made of a flexible fabric that works for an observer in any position. Such a garment can't any longer rely on a passive invisibility that reroutes the incident light; it has to be an 'intelligent', active material that constantly adjusts itself to the circumstances. It is not so much an invisible cloth as a wearable computer.

This means that, for all their cleverness, metamaterials won't suffice.

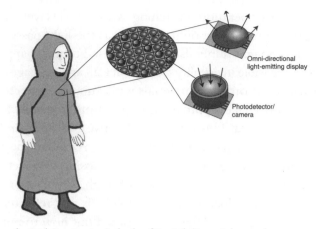

How a hypothetical 'projection' cloak of invisibility might work.

We need to go back to the 'projection cloaks' mentioned in the previous chapter, in which banks of cameras capture, for every viewing angle, the scene that the cloak is obscuring, and then light-emitting diodes reproduce that view on the opposite side, perfectly matched for colour and brightness. The LEDs would be shaped like fish-eye lenses that can project hemispheres of rays, each ray tailored to an observer in the corresponding direction. The lenses would be interspersed with light sensors that detect the scene, while microprocessors within the cloak calculate what output is required from each lens-shaped pixel. That's an immense amount of logic circuitry, and it would demand a substantial power supply to allow each pixel to reproduce the brightness of a sunlit scene. The cameras have to be fast enough to enable the lenses to adjust almost instantaneously to quick-moving objects, and the resolution of the cloak might have to accommodate every leaf and blade of grass.

Wiring up such a complex system would be a nightmare. But Zambonelli and Mamei propose to do away with wiring altogether. Instead, the tiny devices would be autonomous, carrying their own power source (perhaps solar cells) and communicating via wireless radio transmission. The devices could then be positioned randomly all over the surface of the cloak. If they are small enough, they could even be stuck on by spray-painting an optically active version of the self-organizing 'smart dust' that we saw Zambonelli contemplating earlier (see page 226). The researchers calculate that, to create a seamless, sharp image of the background from a distance of about

3.5 metres, these devices would need to be about 3 millimetres across, so that each square metre of cloak holds just over 100,000 of them. Making structures like this, the researchers say, is currently within the realms of technological possibility.

One key problem is how each unit determines its own position within the cloak and selects the other units with which it must communicate. This is a challenge that Zambonelli and Mamei, specialists in 'pervasive computing', are well placed to evaluate. Although the information-processing demands are clear enough for a solid, flat wall seen from a fixed perspective, they become horribly complicated once you consider a curved, three-dimensional cloak that should bc invisible from any vantage point. But it's not impossible in principle, and Zambonelli and Mamei say that internet protocols already tackle similar data-routing challenges within an unstructured and ever-changing ecosystem of communicating devices. All the same, they admit that getting all those multi-directional input and output units integrated side by side on the cloak's surface is stretching the limits of both the hardware and software.

And this is still just for a rigid cloak. As well as hiding military vehicles – a more sophisticated version of the 'invisible tank' being developed by BAE (pages 252–3) – Zambonelli and Mamei suggest some ingenious uses for this kind of system. Virtual windows might be spray-painted onto buildings: from inside, one would see a big, open, airy view of the world outside, while from outside the walls would look just the same as before (a boon for gloomy protected buildings). Imagine how a driver's all-around visibility could be enhanced in a car rendered 'internally transparent' this way (or perhaps don't – it sounds as much unsettling and distracting as safety-enhancing).

But what about a cloak that is flexible and constantly moving? In part this 'just' becomes a question of each optical unit having to regularly update its coordinates. If that's too much for the tiny minds of such devices, perhaps one small part of the cloak could be held fixed so as to provide a constant reference point to which the other units could compare their grid coordinates. Zambonelli and Mamei don't claim that it's possible to make these units yet – but it's possible at least to envisage the kinds of technologies that might be required. This gives the pair licence – it's a gesture of considerable chutzpah – to estimate the cost of the full invisibility cloak. At 'well below a half-million euros', it lands well within reach of the budgets of Google or the US Army.

It is a wonderful, laudable yet quixotic vision. When the Italian duo presented it in a paper, other researchers received the proposal with bemusement. 'These authors have an overactive imagination', said Steve Shafer, a researcher in ubiquitous computing at Microsoft. 'Do I believe we're going to invent invisibility cloaks soon? No. Is this paper to be taken seriously? Not by me.'

But that wasn't meant in scorn. 'It would be kind of cool, wouldn't it?', Shafer added:

> There's something to be said for papers that go way outside the bounds and inspire people to wild feats of imagination . . . If I were teaching, I might give it to my class anyway and ask them to think about it. This might not be possible in my lifetime, but could be in theirs.

Martial Hebert, a specialist in computer vision and robotics at Carnegie Mellon University in Pittsburgh, was rather less taken with the idea. 'It would be wise to motivate work on sensor and processor networks without resorting to Harry Potter', he sniffed. More seriously, Hebert argued that it couldn't work even in principle. For example, because of parallax effects (the perception of distance that we get from binocular vision), 'you cannot reconstruct exactly the view that an observer in an arbitrary position would see.' And even if you could somehow build such effects into the sensor and emitter network, there are bound to be things that an observer can see but a sensor – in a different position – cannot. And that's even before you start to worry about the obstruction caused by folds in the cloak. 'Shuffling the set of input rays into a set of output rays will not generally generate a correct image', Hebert says. The problem is made worse by reflective surfaces in the surroundings, since what gets reflected to a viewer depends on the viewer's position and so can't be properly matched by what is recorded by a light sensor in a different position. In short, creating invisibility by reconstructing the view occluded to an observer can be achieved in general only if you can record that view from *exactly the same position* as the observer.

In one sense this is a technical problem – albeit potentially an insurmountable one, at least for this record-and-project approach. But viewed within the context of the history of invisibility, it takes on another connotation. It tells us that invisibility is not an objective but a subjective issue. It is a property not of a particular fabric but of the onlooker's vision. And so it won't be achieved by manipulating light in some abstract space, but demands that we tamper with what *each individual* sees. Invisibility is in that sense like beauty: in the eye of the beholder.

Myth and meaning

Transformation optics and wireless microtechnologies allow us to imagine tricks seemingly drawn from the farthest reaches of science fiction or natural magic (the place is much the same), of which traditional invisibility is just one. Yet it can be hard to fit the practical possibilities to our mythical templates. Do we really believe that a piece of micromachined silicon or a grid of circuit boards can connect to the stories of Plato and Homer? Is the hair-trigger manipulation of photons in optical fibres really editing history, taking scissors to time?

Evidently there is a wide and perhaps unbridgeable gap between the mythical narrative used to frame scientific discoveries and the limitations of putting them into practice. That creates a rhetorical tension that scientists must constantly confront. Phenomena that seem to them to be equivalent *in principle* – making an object invisible to the eye, say, and diverting microwaves in two dimensions – will strike the outsider as totally different things: one astounding, the other (perhaps quite literally) opaque. The question is not so much about the legitimacy of using popular, familiar stories to excavate and illustrate the underlying character of a scientific innovation – this seems a valuable, even indispensable practice. The issue is rather that what scientists often take as a matter of *technique* is actually a matter of *metaphor*. The elision of the two can become uncomfortable, as H. G. Wells discovered. When science claims to work magic, it had better be prepared for what magic really means. And much of that meaning resides with the motivation: magic is not an open-ended technology, but a means to a specific end.

It should be possible, using banks of sensors and light sources, metamaterials, fancy prisms or smart camouflage, to make some objects hard to see. In ideal situations – a flat surface, say – it might even be possible to render them more or less invisible. And for many technological applications, that could be enough to be useful. It is certainly rather clever and wonderful in its own right. But this is illusionism, a legacy of the music-hall magic of the Vanishing Lady and the Headless Man. It is likely to be cumbersome, difficult, compromised. It is not the invisibility of Plato or Harry Potter, or even of H. G. Wells. This is not the way to kill the king, to save the princess, or to become an invincible tyrant.

Yet we have seen how inexorably the mythic and magical connotations of invisibility attach themselves even to the most limited of real-world realizations. We interpret the science and technology through a mythical lens, no matter how great a gulf separates the imaginary from the actual. Scientists constantly encourage this, and there isn't necessarily any harm in it. But there's something to be said for keeping the distinction in view. For invisibility is by no means the only dream that has been used to this end: we also project onto our scientific discoveries and technological inventions fantasies about immortality and resurrection, telepathy, time travel, space travel, teleportation, anthropoeia (making people), alien life and alchemy, to name just a few. We do this for good reason: myths, including modern and science-fictional ones, enable us to explore the hopes and fears invoked by what might be achievable, and motivate us to make it achievable. But as the history of invisibility shows, myth is no blueprint for the engineer. It is more important than that.

Notes

Chapter 1

Epigraphs
'It seemed that the ring he had': J. R. R. Tolkein (1966), *The Hobbit*, p.79. Unwin & Allen, London.
'And perhaps in this is the whole difference': J. Conrad (1899/1973), *Heart of Darkness*, p.101. Penguin, London.

'committed adultery with the king's wife' Plato (4th century BC/2000), p.36.
'would be so incorruptible': ibid.
'The man who did not wish': ibid., p.37.
'those who practice justice do so': ibid., p.35.
'they threw their arms about each other': R. B. Bottigheimer (2010), p.100.
'cloaks and rings of invisibility are used mostly to enter': F. Vaz da Silva, personal communication.
'in which they believe the self': Russell et al. (2012), p.550.

Chapter 2

Epigraphs
'Then charm me, that I': C. Marlowe (1604/1994), *Dr Faustus*, p.32. Dover, Mineola, NY.
'To render oneself invisible is a very easy matter': S. Liddell Macgregor Mathers (1932), p.147. De Laurence, Chicago.

'Take on Midsummer night': J. Aubrey (1972), p.253.
'Take a black cat, and a new pot': *The Grand Grimoire*, available at http:// www.hermetics.org/pdf/grimoire/The Grand Grimoire - Dark Lodge version.pdf.
'Athal, Bathel, Nothe, Jhoram, Asey': E. Grillot de Givry (1971), pp.183–4.
'perform this work as you all know how': ibid.

'you will make yourself totally hidden': J. M. Greer & C. Warnock (2010–11), p.244.

'The ring must be made of fixed mercury': E. Grillot de Givry (1971), p.185.

'Those itinerants specialized in accessing the invisible': B. M. Stafford & F. Terpak (2001), p.49.

'From vapours and fumes demons': L. Thorndike (1934), Vol. IV, p.304.

'Manie times witches are seene in the fields': R. Scot (1972), pp.43–4.

'You shall read in the legend': ibid., p.45.

'the practical art of natural knowledge': R. Ralley (2010), p.33.

'In this art of naturall magicke': R. Scot (1972), pp.164–5.

'it will hang in the air and tremble': A. Still (1946), p.7.

'Iron and the lodestone are drawn together and united': ibid., p.3.

'that amber does exhale something peculiar': A. Still (1944), p.26.

'All things must be enchained': Plotinus, *Enneads*, available at http://classics. mit.edu/Plotinus/enneads.2.second.html.

'There is therefore a wonderfull vertue': C. Agrippa (1651), *Three Books of Occult Philosophy* Book I, Ch. XIII. Available at http://www. esotericarchives.com/agrippa/agrippa1.htm.

'All teems with symbol': Plotinus, op. cit.

'Behold the herbs! Their virtues are invisible': C. Wilson (1995), pp.43–4.

'the readiness to relate the unrelated': D. S. Katz (2005), p.8.

'hidden from the senses': R. Kieckhefer (1990), p.12.

'the eye of an ape': I. Marathakis (2007), unpaginated.

'pierce the right eye of a bat': ibid.

'As the eyes of the dead': ibid.

'In the use of this stone': Pliny, *Natural History* Book 37, Ch. 60. Available at http://www.perseus.tufts.edu/hopper/text?doc=Perseus%3Atext%3A 1999.02.0137%3Abook%3D37%3Achapter%3D60.

'There is also another vertue': C. Agrippa, op. cit., Book I, Ch. XXIII. Available at http://www.esotericarchives.com/agrippa/agripp1b.htm.

'The Chimera of the Rosie-Crosse': B. Jonson, 'The Underwood', in C. H. Herford, P. Simpson & E. Simpson (eds) (1947), *Ben Jonson, Vol. 8: The Poems; The Prose Works*, p.206. Oxford University Press, Oxford.

'whether they had not any kind of Jewell': F. Godwin (1638), *Man in the Moone*, pp.100–1. John Norton, London.

'Take the stone *Ophthalmus*': I. Marathakis (2007).

'Having placed the ring on a palette-shaped plate': A. E. Waite (1913), *The Book of Ceremonial Magic*, Ch. 7, Sect. 3. Available at http://www.sacred-texts.com/grim/bcm/bcm73.htm.

'quality . . . manifest or occult': B. Copenhaver (1998), p.458.

'These qualities are occult only to the ignorant': ibid., p.467.

'a gentleman . . . who thickens the air': ibid., p.469.

'There are no qualities which are so occult': C. Wilson (1995), p.56.

'tiny invisible instruments': B. Copenhaver (1998), p.471.

'fabricat[ing] faculties or occult qualities': ibid., pp.498–9.

'a substance incorporeal': D. Garber (2006), p.60.

'almost certainly, when readers encountered the phrase': P. Harrison (2011), p.31.

'These Principles I consider not as occult Qualities': D. Garber (2006), p.67.

''Tis utterly inconceivable, that inanimate brute Matter': A. R. Hall (2006), *Philosophers at War*, p.156. Cambridge University Press, Cambridge.

'perhaps the block of marble itself': C. Wilson (1995), p.208.

'gave the occult philosophy a last hour': B. Copenhaver (1998), p.503.

'We, the deputies of the principal College': C. Mackay (1841/2004), *Extraordinary Popular Delusions and the Madness of Crowds*, p.157. Barnes & Noble, New York.

'The frightful Compacts entered into': A. E. Waite (1887), *The Real History of the Rosicrucians*, p.391. George Redway, London. Available at http://www.sacred-texts.com/sro/rhr/rhr29.htm.

'All those who seek entrance into our society': ibid., pp.388–9.

'assert that they can become invisible at pleasure': ibid., p.391.

'they frequent the Sabbaths': ibid., p.397.

'do now and then honour me': C. Webster (1974), p.20.

'to teach a Masonical Art, by which any man': R. Berman (2013), *Schism: The Battle That Forged Freemasonry*, p.43. Sussex Academic Press, Eastbourne.

'I am certain that there must be sciences': W. Chaucer (1968), *The Canterbury Tales*, p.438. Penguin, Harmondsworth.

'by illusion or apparition': ibid., p.441.

'always acknowledge wherein': O. Davies (2012), p.56.

'Both the instructor and the mountebank': B. M. Stafford (1994), p.73.

'to expose the Practices of artful Impostors': R. Ralley (2010), p.79.

'to convince superstition of her many ridiculous errors': C. Pinchbeck (1805), *The Expositor; or, Many Mysteries Unravelled*, p.5. Printed for the author, Boston.

'resembled the charlatans Franz Anton Mesmer': B. M. Stafford & F. Terpak (2001), p.85.

'experiments in natural magic': A. A. Hopkins (1898/1976), p.2.

'Science has laughed away sorcery': ibid., p.6.

'this may be due to the bad fortune': ibid., p.50.

'are now competing with the most noted mediums': ibid., p.96.

'We have, therefore, only the trouble': ibid, pp.96–7.

'I cannot get it out of my mind': C. McIntosh (1980), p.9.

'a delicate, fiery, as it were': Anon. (1852), 'New discoveries in ghosts', *International Magazine of Literature, Art and Science*, **5(3)**, 387–390, here 390.

'I should call [*vril*] electricity': E. Bulwer-Lytton (1871), *The Coming Race*, p.47. William Blackwood, Edinburgh.

'mesmeric power': A. A. Hopkins (1898/1976), p.17.

'men ought to be cautious and to be fully assured': J. Webster (1677), *The Displaying of Supposed Witchcraft*, p.265. J. M., London.

'a natural and divine': E. Levi (1860/1922), p.16.

'warms, illuminates, magnetises, attracts': E. Levi (1860/1922), p.17.

'Come to me, O shroud of darkness': S. Richards (1982), p.136.

'the use of the Will to guide the powers': R. Ralley (2010), p.110.

'the list would be virtually indistinguishable': J. Kane (1995), 'Varieties of mystical experience in the writings of Virginia Woolf', *Twentieth Century Literature* 41(4), 328–49.

'one of the greatest spiritual movements': W. Kandinsky (1912/1946), *On the Spiritual in Art*, p.25. Solomon R. Guggenheim Foundation, New York.

'the means of approaching the unknown': M. E. Warlick (2001), *Max Ernst and Alchemy*, p.171. University of Texas Press, Austin.

'died the 1st of August 1914': ibid., p.17.

'use the most precise methods': G. Parkinson (2008), p.112.

'one of the most pernicious delusions': L. E. Sullivan (ed.) (1987), p.87.

'bastard sister of science': O. Davies (2007), p.17.

'raises the psychological self-confidence': L. E. Sullivan (ed.) (1987), p.92.

'ritualizes man's optimism': ibid.

'the embodiment of the sublime folly': O. Davies (2012), p.63.

Chapter 3

Epigraphs

'Of all the *Arcana* of the invisible World': Daniel Defoe (1729), *The Secrets of the Invisible World Disclos'd*, p.1. J. Clarke et al., London.

'Insensibly, we yielded to the occult force': F.-J. O'Brien (1859).

'As no Writer in any Age': Anon. (1722), *St. James Chronicle*, February 20–22. See http://www.folger.edu/template.cfm?cid=1421.

'His whole demeanour is so expressive of terror': G. C. Lichtenberg (1938), *Lichtenberg's Visits to England as Described in His Letters and Diaries*, trans. M. L. Mare & W. H. Quarrell, pp.9–10. Clarendon Press, Oxford.

'his over-fondness for extravagant attitudes': T. Cibber (1756), *Dissertations on Theatrical Subjects*, pp.28–9. Printed for the author, London.

'Shakespeare managed . . . to lift the whole ghost-business': J. D. Wilson (1967), p.57.

'the linchpin of *Hamlet*': ibid., p.52.

'There is a whole genre of Elizabethan': K. Thomas (1991), p.705.

'Englishmen were seriously aware': R. H. West (1969), p.2.

'distinguish calmly between natural happenings': N. Taillepied (1600/undated), p.xvii.

'Melancholicks and those suffering': ibid., p.12.

'Timorous and fearful Men': ibid., p.22.

'Sometimes a ghost will appear': ibid., p.78.

'the clear unclouded vision of a child': ibid., p.97.

'Very often the servants in a house': ibid., p.79.

'only clothe themselves with a body': ibid., p.112.

'It is now dead midnight': W. Shakespeare, *Richard III*, Act 5, Sc. iii.

'This spirit, dumb to us, will speak to': *Hamlet*, Act 1, Sc. i.

'Between our Ancestors laying too much': D. Defoe (1729), *The Secrets of the Invisible World Disclos'd*, p.1. J. Clarke et al., London.

'personified men's hopes and fears': K. Thomas (1991), p.717.

'Thou art a lying spirit': D. Grant (1965), p.29.

'the opinion of the whole assembly': S. Johnson (1762), *Gentleman's Magazine*, February (abr. Boswell).

'one supreme ghost': K. Thomas (1991), p.706.

'The suspicion is solidified': V.-M. Kärkkäinen (2002), p.135.

'it is conceivable that he': J. McIntyre (1997), p.265.

'the life force immanent in all the living': J. Moltmann (1992), *The Spirit of Life*, pp.225–6. Fortress Press, Minneapolis.

'Could it be that the most compelling response': V.-M. Kärkkäinen (2002), p.161.

'intelligent Creature[s] of the invisible World': O. Davies (2007), p.109.

'[They] eat and enjoy the product of their labour': H. E. Sigerist (1941), *Paracelsus: Four Treatises*, p.230. Johns Hopkins Press, Baltimore.

'woo man [and] seek him assiduously': ibid., p.239.

'Some are in appearance like the gnomes': A. Conan Doyle (1922), p.149.

'She, that pinches country wenches': K. Thomas (1991), p.731.

'knew that he could never count on actually *seeing* the fairies': ibid., p.733.

'so quick . . . that none but those for whom': ibid.

'It is frightfully difficult to know much about the fairies': J. M. Barrie (1913), *The Little White Bird*, p.157. Charles Scribner's Sons, New York.

'Mr. Tom Charman, who builds for himself': A. Conan Doyle (1922), p.140.

'for the little folks to consume': ibid., p.144.

'and especially in Wales and Ireland': ibid.

'one of these Great Ones has a new idea': ibid., p.187.

'Pod looked at her blankly': M. Norton (1998), *The Borrowers*, p.28. Harcourt, New York.

'If they are conventional': A. Conan Doyle (1922), p.56.

'Short of final and absolute proof': ibid., pp.39–40.

'The cry of "fake" is sure to be raised': ibid., p.40.

'There are few realities which cannot': ibid., p.6.

'We see objects within the limits which make up': ibid., p.14.

'It seems to me that with fuller knowledge': ibid., p.55.

'In the field we saw figures': ibid., pp.108–14.

'The recognition of [the fairies'] existence': ibid., p.58.

'Two village kids and a brilliant man': 'Fairies, Phantoms, and Fantastic Photographs', *Arthur C. Clarke's World of Strange Powers*, ITV, 22 May 1985.

'The photograph is a farewell': E. Cadava (1997), *Words of Light: Theses on the Photography of History*, p.13. Princeton University Press, Princeton.

'the visible, the material invisible': J. Coates (1911), p.1.

'To say that the invisible cannot be photographed': ibid., p.2.

'the odic force, the ether': E. Schenkel & S. Welz (eds) (2007), p.134.

'By the artificialness of some': O. Davies (2007), p.187.

'pictures on glass, to make strange things': S. Pepys, Diary, Sunday 19 August 1666. Available at http://www.pepysdiary.com/diary/1666/08/19/.

'In the seventeenth century': O. Davies (2007), p.191.

'scientific necromancy . . . unholy marriage': B. M. Stafford & F. Terpak (2001), pp. 88, 85.

'I shall drive the audience out': O. Davies (2007), p.27.

'I reached a point where my physical reflection': A. Crowley (1969), *The Confessions of Aleister Crowley: An Autohagiography*, ed. J. Symonds & K. Grant, p.204. Jonathan Cape, London.

'I was able to take a walk in the street': ibid.

'The cinematic medium perfected': M. Solomon (2010), p.3.

'have found a way to revive the dead': ibid., pp.11–12.

'the art of ghosts, a battle of phantoms': K. McMullen (1983).

'[i]t seems as if these people': R. Clarke (2013), p.274.

'the main reason for the disappearance': K. Thomas (1971), p.723.

'distorted, uncouth, and horrible': F.-J. O'Brien (1859).

'I reckon that if there were such a thing': O. Wilde (1887/1961), 'The Canterville Ghost', in G. F. Maine (eds), *The Works of Oscar Wilde*, p.193. Collins, London.

'There is many a mother mourning': G. Wood (2002), *Living Dolls*, p.123. Faber, London.

'Death has lost some of its sting': J. D. Peters (1999), p.142.

'may be expected to find that the sound waves': C. S. Peirce, C. Hartshorne,

P. Weiss (eds) (1935), *Collected Papers of Charles Sanders Peirce*, Vols I–IV, p.383. Harvard University Press, Cambridge, MA.

'By preserving people's apparition': J. D. Peters (1999), p.139.

'I believe that ghosts are a part of the future': K. McMullen (1983).

'To interact with another person': J. D. Peters (1999), p.142.

'Written kisses don't reach their destination': ibid., epigraphs.

'The spirits won't starve': ibid.

'The concern in psychical research': ibid., p.141.

Chapter 4

Epigraphs

'Invisible things are the only realities': W. Godwin (1817), *Mandeville*, Vol. III, p.48. Archibald Constable & Co., Edinburgh.

'Is the invisible visible?': O. Glasser (1933), p.8.

'The vast interplanetary and interstellar regions': J. C. Maxwell (1876), *Scientific Papers of James Clerk Maxwell*, Vol. 2, LIV, p.311.

'One thing we are sure of': S. P. Thompson (1910), *The Life of William Thomson, Baron Kelvin of Largs*, p.1035. Macmillan, London.

'The electromagnetic fields appear as ultimate': A. Einstein (1920), 'Ether and the Theory of Relativity', talk on 5 May at the University of Leiden (in German). Published in A. Einstein, *Sidelights on Relativity*, trans. G. B. Jeffrey & W. Perrett, pp.3–24. Methuen, London, 1922. Available at http:// www-history.mcs.st-andrews.ac.uk/Extras/Einstein_ether.html.

'There is a weighty argument': ibid.

'We know little about the medium': S. Quinn (1995), p.207.

'At an early age': O. Lodge (1931), *Past Years*, p.111. Hodder & Stoughton, London.

'teems with theories in the nature': J. B. Stallo (1881), p.128.

'the power is gone; and this only because': F. A. J. L. James (ed.) (1999), *The Correspondence of Michael Faraday*, Vol.4, p.527. Institution of Electrical Engineers, Stevenage.

'My sister Katie was the first to observe': H. Houdini (1924/1972), *A Magician Among the Spirits*, pp.7–8. Arno Press, New York.

'telegraphy has been until lately an art occult': Editorial (1862), *The Electrician* **2**, 39 (May).

'I am working on the theory': J. Sconce (2000), p.81.

'excite wonder at the marvellous feats': R. Noakes (1999), p.425.

'in the position of an electrician at Valentia': ibid., p.443.

'There are probably other powers': ibid.

'Whether vibrations of the ether': W. Crookes (1892), 'Some possibilities of electricity', *Fortnightly Review* **51**, 175.

'Why, in fact, if one wire can talk to another': S. Natale (2011), p.267.

'mental telepathy is not a jest': ibid., p.266.

'We have been toying with the intangible': J. Sconce (2000), p.66.

'Americans began to realize that they were all': ibid., p.69.

'Have you ever seen a spiritualistic seance?': R. Kipling (1920), p.220.

'We are only telephone wires': M. Cohen (ed.) (1965), *Rudyard Kipling to Rider Haggard, the Record of a Friendship*, p.100. Hutchinson, London.

'leaving the professional reader and time': W. Brock (2005), p.59.

'Something new and worthy of the notice': ibid., p.126.

'At first we had rough manifestations': ibid., pp.137–8.

'is himself subject to unaccountable': G. Stein (1993), p.98.

'certain of Home's phenomena fall quite outside': W. Brock (2005), p.139.

'All I am satisfied of is that there exist': ibid., p.129.

'Assuming that there are invisible beings': ibid., p.128.

'These experiments *confirm beyond doubt*': ibid., p.147.

'science has made a hole': ibid., p.176.

'While Crookes, arguably, would still have': ibid., p.209.

'When the radiometer was first constructed': ibid., p.211.

'At Gabriel College there was a very holy object': P. Pullman (1995), *The Northern Lights*, p.149. Scholastic, London.

'we seem at length to have within our grasp': W. Crookes (1879a), 439.

'as far removed from those': W. Brock (2005), p.231.

'If we conceive a change as far beyond': W. Crookes (1879a), 419.

'The phenomena in these exhausted tubes': W. Crookes (1879b), 164.

'We have actually touched the border-land': W. Crookes (1879a), pp.439–440.

'etherial or astral body': R. Noakes (2008), p.329.

'asked her permission to clasp her': W. Brock (2005), p.186.

'I had been led by the accounts of witnesses': T. H. Hall (1962), p.22.

'Photography is as inadequate to depict': R. G. Medhurst & K. M. Goldney (eds) (1972), p.139.

'was thus the impresario who chose': T. H. Hall (1962), p.105.

'series of bewildering effects': W. Brock (1995), p.195.

'Yet these ultimate particles': W. Crookes (1874), *Researches in the Phenomena of Spiritualism*, p.33. J. Burns, London.

'He believed, in common with so many other dupes': G. Stein (1993), p.11.

'would compare favourably with the lower': C. W. Leadbeater (1942), p.17.

'looking for invisible rays': O. Glasser (1933), p.13.

'If we imagine other creatures of such infinite': A. Hessenbruch (2000), p.105.

'the last new mystery that human genius': K. Williams (2007), p.50.

'so completely and irresistibly': S. Natale (2011), p.265.

'go away and photograph Mahatmas': *Punch* 25 January 1896, in O. Glasser (1933), p.41.

'a cheap piece of metal': ibid., p.205.

'The Roentgen Rays': ibid., p.44.

'Nobody knows what other invisible pencils': E. E. Slosson (1920), 'Sun dogs', *Independent*, 2 October, p.10.

'Imagine that someone wants to know': S. Natale (2011), p.270.

'It emerged suddenly out of nowhere': M. Malley (2011), p.209.

'The glowing tubes looked like faint fairy lights': M. Curie (1923), *Pierre Curie*, trans. C. & V. Kellogg, p.187. Macmillan, New York.

'Here the question can be raised whether mankind': P. Curie (1905), 'Radioactive substances, especially radium', 1903 Nobel Prize in Physics lecture. Available at http://www.nobelprize.org/nobel_prizes/physics/laureates/1903/pierre-curie-lecture.pdf.

'Chemists were being asked to accept': M. Malley (2011), p.29.

'the disconcerting character of the new chemistry': ibid., p.65.

'Whatever your Ill, write us': ibid., p.212.

'It possessed, said Dr Bailey': *Star*, 7 May 1909, p.2. Available at http://paperspast.natlib.govt.nz/cgi-bin/paperspast?a=d&d=TS19090507.2.29.

'albeit a very persistent one': the putative source of this widely attributed quotation is not clear.

'Around the turn of the century': M. Malley (2011), p.201.

Chapter 5

Epigraphs

'It is not a very incredible thing to suppose': W. F. Barrett (1917), p.112.

'It is not difficult to design thought-experiments': F. Wilczek (2013), p.1.

'sensitive, nervous person': R. Noakes (2004), p.427.

'unseen connection . . . so very magical': ibid., p.429.

'more appropriate for a conjuror's stage': ibid.

'forces unrecognized by our senses': ibid., p.435.

'intelligence, thought and will': W. Brock (2005), p.355.

'human-like, but not really human': W. F. Barrett (1917), p.113.

'the passage of matter through matter': ibid., p.114.

'invasion of our will': ibid., p.250.

'we cannot reasonably expect to see psychomeres': E. E. Fournier d'Albe (1908), p.147.

'And thus it comes about that all the fairies': ibid., p.184.

'earth memories . . . should be awakened': ibid., p.159.

'first, a fine mist, then a cloud': ibid.

'why draw the line at a little additional': ibid., p.112.

'no higher than the average human being': ibid., p.309.

'theology has been ruthlessly evicted': ibid., p.135.

'We must resolutely combat the tendency': ibid., p.164.

'The invisible force fields of electromagnetism': M. Malley (2011), p.8.

'the power that one mind': B. Stewart & P. G. Tait (1876), p.43.

'Circle-squarers, Perpetual-motionists': R. Noakes (1999), p.448.

'the presumed incompatibility of Science': B. Stewart & P. G. Tait (1876), p.vii.

'intrude on the region of *knowledge*': J. Tyndall (1874), *John Tyndall's Address Delivered Before the British Association Assembled at Belfast, With Additions.* London: Longmans, Green, and Co. Available at http://www.victorianweb.org/science/science_texts/belfast.html.

'The visible universe must': B. Stewart & P. G. Tait (1876), p.64.

'we are forced to believe': ibid., pp.64, 157.

'peculiarity of structure which is handed over': ibid., p.180.

'May we not regard ether': ibid., pp.158–9.

'Certain molecular motions . . . free to exercise its functions': ibid., p.159.

'some existing in different parts of space': ibid., pp.160–1.

'restore energy in the present universe': ibid., p.165.

'The scientific difficulty with regard to miracles': ibid., p.176.

'It appears to us as almost self-evident': ibid., p.190.

'ought to live for the unseen': ibid., p.192.

'unraveling the mysteries of the invisible universe': J. P. Ostriker & S. Mitton (2013), *Heart of Darkness: Unravelling the Mysteries of the Invisible Universe.* Princeton University Press, Princeton.

'Are we human beings': F. Wilczek (2013), p.2.

'in the past scientists': ibid., p.6.

'giant branes right next door': B. Greene (2011), p.116.

'The act of making a decision': M. Tegmark (2003), 'Parallel Universes', *Scientific American*, May, p.48.

'each copy *is* you': B. Greene (2011), p.229.

Chapter 6

Epigraphs

'Q: Did you notice anything': C. F. Chabris & D. J. Simons (2010), *The Invisible Gorilla*, p.6. Crown, New York.

'Sue had said people': C. Priest (1984), p.225.

'is therefore entirely in a power': E. Lévi (1896), p.304.

'assumed a stooping gait': ibid., p.305.

'the true Ring of Gyges is the will': ibid.

'the symbolic sense of the ring': S. Richards (1982), p.72.

'He would pass through a room': V. Lytton (1913), *The Life of Edward Bulwer, by His Grandson, Earl of Lytton*, Vol. 2, p.40. Macmillan, London.

'A modern conjurer who': A. A. Hopkins (1898/1976), p.25.

'From an eye-witness I had it': H. S. Olcott (1895), *Old Diary Leaves: The True Story of the Theosophical Society*, p.23n. G. P. Putnam's & Sons, New York & London.

'it never occurred to me that I was': ibid., pp.46–7.

'nothing but a certain delusion': *Malleus Maleficarum*, Part 1, Question 9. Available at http://www.sacred-texts.com/pag/mm/mm01_09a.htm.

'by the agility of men': ibid.

'If his luminous eyes': A. A. Hopkins (1898/1976), p.22.

'He would place a ring upon the finger': ibid., p.23.

'matter destitute of life': R. Noakes (2008), p.331.

'quite out of place': ibid.

'the supernatural returns as the erotic': D. Katz (2005), p.146.

'one more example of the remarkable fact': ibid., p.148.

'Wishing to see, on account of its medico-legal bearing': B. Sidis (1921), p.109.

'in the name of science': R. Clarke (2013), pp.182–3.

'As male invisibles grow older': C. Priest (1984), p.125.

'The urge to rewrite ourselves': ibid., p.234.

'contented girl, clever at school . . . generalisations and platitudes': ibid., p.201.

'I've been invisible to Mum': ibid., p.212.

'That's the way they account for it': ibid.

'seemd to be totaly engag'd': Diary of Joseph Banks, available at http://southseas.nla.gov.au/journals/banks/17700428.html.

'Stage hypnotists work a similar effect': C. Priest (1984), p.41.

'It is the cloth that captivates us': M. Tatar (2008), p.xxv.

'insisting that the value': H. Robbins (2003), p.663.

'Arguably enchantment is the core': F. Vaz da Silva, personal communication.

'To "see" in a fundamental sense': F. Vaz da Silva (2002), p.44.

Chapter 7

Epigraphs

'I am one of the most irresponsible beings': R. Ellison (1947), p.14.

'The invisible man is a threat': H. G. Wells (1897/2005), p. xvii.

'From the start, the two were as different': F. Karl (1973), 'Conrad, Wells, and the Two Voices', *PMLA* **88**(5), 1049–65, here p.1052. Modern Language Association of America.

'I am always powerfully impressed': P. Holt (1992).

'the strangest appearance conceivable': H. G. Wells (1897/2005), p.7.

'The poor soul's had an accident': ibid., p.8.

'in a big ill-managed lodging-house': ibid., p.94.

'To do such a thing would be to transcend': ibid., p.92.

'This announces the first day': ibid., p.134.

'He may lock himself away': ibid.

'naked and pitiful on the ground': ibid., p.148.

'there ended the strange experiment': ibid.

'Full of secrets': ibid., p.150.

'vast and incredible mouth': ibid., p.11.

'a waving of indecipherable shapes': ibid., p.17.

'They were prepared for scars': ibid., p.37.

'He may be watching me now': ibid., p.135.

'is really almost persuaded': Anon. (1897), *Literature* **1** (30 October), p.50.

'Röntgen Rays and other still mysterious vibrations': ibid.

'For the writer of fantastic stories': H. G. Wells (1933), *The Scientific Romances of H. G. Wells*, p.viii. Victor Gollancz, London.

'in some denser liquid than water': H. G. Wells (1987/2005), pp.90–1.

'the whole fabric of man': ibid., p.91.

'And suddenly, not by design': ibid., p.92.

'a sort of ethereal vibration': ibid., p.95.

'With the right pigments': J. London (1903).

'two little ghosts of her eyes': H. G. Wells (1897/2005), p.96.

'nothing save where an attenuated': ibid., p.100.

'My head was already teeming': ibid., p.102.

'revel in my extraordinary advantage': ibid., p.103.

'At night, when all around is still': P. Haining (ed.) (1978), p.60.

'I went over the heads': Wells, p.109.

'credible figure . . . strange and terrible thing': ibid., pp.114, 115.

'He has cut himself off': ibid., p.111.

'I did not lift a finger to save': ibid., pp.94–5.

'Wells turns against the implications': J. Batchelor (1985), p.22.

'trampled calmly over the child's body': R. L. Stevenson (1903), *The Strange Case of Dr. Jekyll and Mr Hyde, With Other Fables*, p.6. Longmans, Green & Co., London.

'we can lighten the darkness': J. Tyndall (1870), 'Scientific Use of the

Imagination', in *Essays on the Use and Limit of Imagination in Science*, p.16. Longmans, Green & Co., London.

'Linked together by the scientific imagination': S. McLean (2009), p.74.

'make people sit up in the cinema': J. Curtis (1998), p.200.

'Obviously, it was a better idea': ibid.

'If the man had remained sane': ibid., p.127.

'If a man said to you': ibid.

'I observed the wild oats': A. Pierce (1898), 'The Damned Thing'. Available at http://classiclit.about.com/library/bl-etexts/abierce/bl-abierce-damned.htm.

'As I looked at this chair': H. Stephen (1881), 'No. 11 Welham Square'. Available at http://gutenberg.net.au/ebooks06/0606521h.html.

'We realized that there was something': http://www.davidmccallumfansonline.com/invman.htm.

'*The Invisible Man* was really a one-joke show': ibid.

'It occurred to me that the man': R. Ellison (1947), p.4.

'I'm an invisible man and it placed me': ibid., p.559.

'Negro life contains the necessity': M. Klein (1964), *After Alienation*, p.132. World, New York.

'I am invisible, understand, simply because': R. Ellison (1947), p.3.

'It came upon me slowly': ibid., p.562.

'When one is invisible he finds such problems': ibid., p.559.

'Behold the Seen Unseen': ibid., p.484.

'In the first place I became invisible': P. A. Cantor (2010).

'What astonishes and seems to derive': J. Matlock (1996), p.190.

'To be stylish, one must speak in society of nothing': ibid., p.186.

'The Invisible Woman Shows burst into': ibid., p.188.

'run around morning and night': ibid.

'the most flattering compliments': ibid.

'This example should serve as a lesson': ibid., p.187.

'Never had a magic trick been so prominent': K. Beckman (2003), p.52.

'turns on the thrill of imagining': ibid., p.58.

'The triple appearance of the vanishing lady': ibid., p.66.

'hovers endlessly between visible': http://www.dukeupress.edu/Vanishing-Women/.

'the way that the internet can encourage': C. Hardaker (2013), 'What is turning so many young men into internet trolls?', *Guardian* 3 August. Available at http://www.theguardian.com/media/2013/aug/03/how-to-stop-trolls-social-media.

'Computer technology has . . . empower[ed] individuals': C. Stryker (2012), p.14.

'normal, well-adjusted people': http://knowyourmeme.com/memes/ greater-internet-fuckwad-theory.
'comment sections can degenerate': http://www.huffingtonpost.com/ jimmy-soni/why-is-huffpost-ending-an_b_3817979.html.
'I think anonymity on the Internet': C. Stryker (2012), p.11.

Chapter 8

Epigraphs
'For the limits, to which our thoughts': R. Hooke (1665/2007), p.16.
'There may be more intelligence': C. Wilson (1995), p.191.

'I have observed many tiny animals': W. R. Shea & M. Artigas (2003). *Galileo in Rome*, p.121. Oxford University Press, Oxford.
'By the help of Microscopes': R. Hooke (1665/2007), p.8.
'in the hidden and innermost recesses': C. Wilson (1995), p.179.
'the magnetical effluviums of the loadstone': M. Boas Hall (ed.) (1970), *Nature and Nature's Laws*, pp.125–6. Macmillan, London.
'Those effects of Bodies': R. Hooke (1665/2007), p.41.
'For, to speak truly': J. Locke (1690), *Essay Concerning Human Understanding*, Book II, Ch.23:10. Available at http://oregonstate.edu/instruct/phl302/ texts/locke/locke1/Book2b.html#Chapter%20XXIII.
'The infinite wise Contriver of us': ibid., Ch.23:12.
'if that most instructive of our senses': ibid.
'Their skins appeared so coarse': J. Swift (1727/1985), *Gulliver's Travels*, P. Dixon & J. Chalker (eds), p.158. Penguin, London.
'snouts with which they rooted': ibid., p.152.
'very minute and almost invisible creatures': C. Wilson (1995), p.155.
'there might living Creatures be seen': ibid., p.169.
'The worms that breed in Humane Bodies': ibid., p.160.
'pick a hole . . . only a statistical certainty': C. G. Knott (1911), *Life and Scientific Work of Peter Guthrie Tait*, pp.213, 215. Cambridge University Press, Cambridge.
'Call him no more a demon but a valve': ibid., p.215.
'a doorkeeper, very intelligent': P. M. Harman (2001), p.143.
'a being with no preternatural qualities': W. Thomson (1879), 'The sorting demon of Maxwell', *Royal Institution Proceedings* **9**, 113.
'In every *little particle* of matter': R. Hooke (1665/2007), p.17.
'So, naturalists observe': J. Mitford (ed.) (1880), *The Poetical Works of Swift*, p.176. Houghton, Mifflin & Co., Boston.
'Let him see therein an infinity': E. E. Knoebel (ed.) (1988), *Classics of Western*

Thought, 4th edn, p.45. Harcourt Brace Jovanovich, New York.

'eat, and fight, and love, and die': E. E. Fournier d'Albe (1907), p.3.

'without the slightest net effect': ibid., p.69.

'The universe, as it is within [man's] experience': R. Noakes (2008), p.324.

'Here is a book intended to expound': M. J. Nye (1996), *Before Big Science*, pp.69–70. Twayne, New York.

'Residual associations between contagion': R. Porter (1996), p.103.

'Physicians' accounts of the causes': ibid., p.105.

'Where are these little beasts?': J. Waller (2002), p.70.

'The fact that there are germs': T. Fontane (1895/1967), *Effi Briest*, p.79. Penguin, Harmondsworth.

'The microscope has revealed the existence': J. Amato (2000), p.118.

'Their ways are strange': A. Baron (1957), p.13.

'The evil spirits crowd the public places': ibid.

'The Invisible Revealing of the Dangerously Beautiful': https://aaas.confex.com/aaas/2013/webprogram/Session5934.html.

'interaction has once again become': B. M. Stafford & F. Terpak (2001), p.4.

'our anxieties have been flattened': ibid., p.43.

'clouds of sub-millimeter-scale microcomputers': M. Mamei & F. Zambonelli (2004), p.1.

'We could imagine a spray to transform': ibid., p.2.

'Rather, they live in an environment': F. Zambonelli et al. (2005).

'will permit the ghost in the machine': B. M. Stafford & F. Terpak (2001), p.66.

'The typical modern "Enlightened" association': ibid., p.53.

Chapter 9

Epigraphs

'All moved under a cloak of invisibility': N. Rankin (2008), p.13.

'Visible form can only be distinguished': H. B. Cott (1940), p.4.

'If we paint a wall': http://news.bbc.co.uk/1/hi/world/asia-pacific/2777111.stm.

'disappears like magic when it settles': P. Forbes (2009), p.47.

'a much less perfect imitation': ibid., p.47.

'After another long time, what with': R. Kipling (1902), *Just-So Stories*, p.45. Doubleday Page & Co., New York.

'In the thin cover described': H. B. Cott (1940), p.94.

'By the contrast of some tones': ibid., pp.50, 51.

'tend to be interpreted as representing': ibid., p.55.

'God as a professional colleague': P. Forbes (2009), p.74.

'All patterns and colors': T. Roosevelt (1911), 'Revealing and concealing coloration in birds and mammals', *Bulletin of the American Museum of Natural History* **30**, 120–221, here 122–3.

'It is based on the American slang': H. B. Cott (1940), p.47.

'When at the same time the animal': ibid., p.36.

'No discovery in the wide field of animal coloration': E. Poulton & A. H. Thayer (1902), *Nature* **65**, 596.

'the doctrine seems to me to be pushed': T. Roosevelt (1911), op. cit., p.122.

'the obscure mental processes which are responsible': ibid., p.229.

'no color scheme whatever': P. Forbes (2009), p.80.

'only in the brightest moments': ibid., p.93.

'To be invisible to the enemy': N. Rankin (2008), p.44.

'to destroy completely the continuity': P. Forbes (2009), p.87.

'The idea is not to render the ship': Rankin, p.130.

'paint a ship with large patches': H. Murphy & M. Bellamy (2009), p.182.

'When I got over there': P. Forbes (2009), p.93.

'How the hell do you expect me': ibid., p.94.

'of course has a relation': ibid., p.97.

'for those parts of the ship': ibid.

'As needs must, we painted sections': D. W. Bone (1919), *Merchantmen-at-Arms: The British Merchant Service in the War*, p.164. Chatto & Windus, London.

'A convoy of heavy guns': R. Penrose (1981), p.199.

'If they want to make an army invisible': ibid.

'I was very happy': P. Forbes (2009), p.104.

'bring down ridicule upon the art': H. B. Cott, 'Camouflage in modern warfare', *Nature* **145**, 949 (1940).

'they are not painted for deception at close range': H. B. Cott (1940), pp.53–4.

'placed at the disposal of the War Office': J. Maskelyne (1949), p.13.

'in connection with the flight': ibid.

'provided several magicians to assist': ibid., p.68.

'For six weeks I had to attend lectures': ibid., p.17

'A lifetime of hiding things on the stage': ibid.

'Magic helped to save Britain': ibid., p.13.

'For years, I and others made': ibid., p.14.

'I took care to remain "invisible"': ibid.

'a special kind of black felt powder': ibid., p.176.

'It sounds more like a scene from a Harry Potter movie': http://www.baesystems.com/magazine/BAES_019786/adaptiv--a-cloak-of-invisibility.

'If in fact the Navy did somehow succeed': C. Berlitz & W. Moore (1979), p.59.

'Records in the Operational Archives Branch': http://www.history.navy.mil/faqs/faq21-1.htm.
'It could be said that degaussing': ibid.

Chapter 10

Epigraphs
'Visible bodies may be made invisible': F. Hartmann (1896), *The Life of Philippus Theophrastus Bombast of Hohenheim, Known By the Name of Paracelsus, and the Substance of His Teachings*, p.294. Kegan Paul, Trench, Trubner & Co., London.
'Imagine there were no practical limits': U. Leonhardt & T. Philbin (2009), p.69.

'we were met with puzzled looks': W. Cai & V. Shalaev (2011), p.30.
'with the wires that could control permittivity': D. R. Smith (undated).
'Every time I read through this paper': ibid.
'why would the general public be the least bit': ibid.
'As the time of the APS meeting': ibid.
'rather like taking a pin and pushing it': D. R. Smith, personal communication. See also P. Ball (2007).
'I thought what could be the next': U. Leonhardt, personal communication.
'The central idea of Pendry's': ibid.
'Researchers announced today': S. Markey (2006), 'First invisibility cloak tested successfully, scientists say', *National Geographic News* 19 October. Available at http://news.nationalgeographic.com/news/2006/10/061019-invisible-cloak.html.
'It is a bit like magic': P. Ball (2008), 'Opening the door to Hogwart's', *Nature* online news, 19 September. Available at http://www.nature.com/news/2008/080919/full/news.2008.1113.html.
'well below a half-million euros': F. Zambonelli & M. Mamei (2002), p.69.
'These authors have an overactive imagination': ibid., p.65.
'There's something to be said for papers': ibid.
'It would be wise to motivate work': ibid., p.64.
'you cannot reconstruct exactly the view': ibid.
'Shuffling the set of input rays': ibid.

Bibliography

J. A. Amato (2000). *Dust: A History of the Small and the Invisible*. University of California Press, Berkeley, CA.

Anon. (2012). *Invisible: Art About the Unseen, 1957-2012*. Hayward Publishing, London.

A. Assmus (1995). 'Early history of X-rays', *Beam Line* Summer, 10–24.

J. Aubrey (1972). *Three Prose Works*, ed. J. Buchanan-Brown. Centaur Press, Fontwell.

N. Baker (1994). *The Fermata*. Vintage, London.

P. Ball (2007). 'TR10: invisible revolution', *Technology Review* March/April. Available at http://www2.technologyreview.com/article/407474/tr10-invisible-revolution/.

A. L. Baron (1957). *Man Against Germs*. E. P. Dutton & Co., New York.

W. F. Barrett (1881). 'Mind-Reading versus Muscle-Reading', *Nature* **24**, 212.

W. F. Barrett (1917). *On the Threshold of the Unseen*. Kegan Paul, Trench, Trubner & Co., London.

J. Batchelor (1985). *H. G. Wells*. Cambridge University Press, Cambridge.

K. Beckman (2003). *Vanishing Women: Magic, Film and Feminism*. Duke University Press, Durham.

B. Bergonzi (1961). *The Early H. G. Wells*. Manchester University Press, Manchester.

C. Berlitz & W. Moore (1979). *The Philadelphia Experiment: Project Invisibility*. Souvenir Press, London.

A. Besant & C. W. Leadbeater (1919). *Occult Chemistry*, revised edn. Theosophical Publishing House, London.

R. B. Bottigheimer (2010). *Fairy Tales: A New History*. SUNY Press, New York.

W. H. Brock (2008). *William Crookes (1832–1919) and the Commercialization of Science*. Ashgate, Aldershot.

E. Bulwer-Lytton (1842; 1971). *Zanoni: A Rosicrucian Tale*. Steiner, Blauvelt, NY, 1971.

W. Cai & V. Shalaev (2011). 'Into the visible', *Physics World* **24**(7), 30–34.

C. Canfield (1989). *Multiple Exposures: Chronicles of the Radiation Age*. Penguin, Harmondsworth.

P. A. Cantor (2010). 'The Invisible Man and the Invisible Hand: H. G. Wells's critique of capitalism', in P. A. Cantor & S. Cox (eds), *Literature and the Economics of Liberty*. Ludwig von Mises Institute, Auburn, AL.

J. J. Cerullo (1982). *The Secularization of the Soul*. ISHI, Philadelphia.

H. Chen & C. T. Chan (2007). 'Acoustic cloaking in three dimensions using acoustic metamaterials', *Applied Physics Letters* **91**, 183518.

H. Chen, X. Luo, H. Ma & C. T. Chan (2008). 'The anti-cloak', *Optics Express* **16**, 14603–14608.

H. Chen, R.-X. Miao & M. Li (2010). 'Transformation optics that mimics the system outside a Schwarzchild black hole', *Optics Express* **18**, 15183–15188.

H. Chen, B. Zheng, L. Shen, H. Wang, X. Zhang, N. I. Zheludev & B. Zhang (2013). 'Ray-optics cloaking devices for large objects in incoherent natural light', *Nature Communications* **4**, 2652.

X. Chen, Y. Luo, J. Zhang, K. Jiang, J. B. Pendry & S. Zhang (2011). 'Macroscopic invisibility cloaking of visible light', *Nature Communications* **2**, 176.

R. Clarke (2013). *A Natural History of Ghosts*. Penguin, London.

J. Coates (1911). *Photographing the Invisible: Practical Studies in Spirit Photography, Spirit Portraiture, and Other Rare but Allied Phenomena*. L. N. Fowler, London.

A. Conan Doyle (1922). *The Coming of the Fairies*. G. H. & Co., New York.

B. Copenhaver (1998). 'The occult tradition and its critics', in D. Garber & M. Ayers (eds), *The Cambridge History of Seventeenth Century Philosophy*. Cambridge University Press, Cambridge.

B. P. Copenhaver (2006). 'Magic', in K. Park & L. Daston (eds), *The Cambridge History of Science Volume 3: Early Modern Science*, pp.518–540. Cambridge University Press, Cambridge.

H. B. Cott (1940). *Adaptive Coloration in Animals*. Methuen, London.

W. Crookes (1879a). 'On radiant matter', *Nature* **20**, 419–440.

W. Crookes (1879b) 'On the illumination of lines of molecular pressure, and the trajectory of molecules', *Philosophical Transactions of the Royal Society* **170**, 135–64.

M. Crosland (1992). *Science Under Control: The French Academy of Sciences 1795–1914*. Cambridge University Press, Cambridge.

J. Curtis (1998). *James Whale: A New World of Gods and Monsters*. Faber & Faber, London.

O. Davies (2007a). *Popular Magic: Cunning-folk in English History*. Continuum, London.

O. Davies (2007b). *The Haunted: A Social History of Ghosts*. Palgrave, Basingstoke.

O. Davies (2012). *Magic: A Very Short Introduction*. Oxford University Press, Oxford.

D. Defoe (1770). *The Secrets of the Invisible World: or A General History of Apparitions*. London.

M. Draper (1987). *H. G. Wells*. Macmillan, Basingstoke.

S. During (2002). *Modern Enchantments: The Cultural Power of Secular Magic*. Harvard University Press, Cambridge, MA.

R. Ellison (1947/1972). *Invisible Man*. Vintage, New York.

V. I. J. Flint (1994). *The Rise of Magic in Early Medieval Europe*. Princeton University Press, Princeton.

P. Forbes (2009). *Dazzled and Deceived: Mimicry and Camouflage*. Yale University Press, New Haven.

I. McL. Forsyth (2012). 'From dazzle to the desert: a cultural-historical geography of camouflage', PhD thesis, University of Glasgow.

E. E. Fournier d'Albe (1907). *Two New Worlds*. Longmans, Green & Co., London.

E. E. Fournier d'Albe (1908). *New Light on Immortality*. Longmans, Green & Co., London.

L. Gamwell (2002). *Exploring the Invisible: Art, Science, and the Spiritual*. Novartis/Princeton University Press, Princeton.

D. Garber (2006). 'Physics and foundations', in K. Park & L. Daston (eds), *The Cambridge History of Science Volume 3: Early Modern Science*, pp.19–69. Cambridge University Press, Cambridge.

B. J. Gibbons (2001). *Spirituality and the Occult: From the Renaissance to the Modern Age*. Routledge, London.

O. Glasser (1933). *Wilhelm Conrad Röntgen and the Early History of the Roentgen Rays*. John Bale, Sons & Danielsson, London.

D. Grant (1965). *The Cock Lane Ghost*. Macmillan, London.

J. Grasset (1910). *Marvels Beyond Science*. Funk & Wagnalls, New York.

S. Greenblatt (2002). *Hamlet in Purgatory*. Princeton University Press, Princeton.

B. Greene (2011). *The Hidden Reality*. Alfred A. Knopf, New York.

J. M. Greer & C. Warnock (eds) (2010–11). *The Complete Picatrix*. Adocentyn Press.

E. Grillot de Givry (1971). *Witchcraft, Magic and Alchemy*, trans. J. C. Locke. Dover, New York.

A. Grove (1997). 'Röntgen's ghosts: photography, X-rays and the Victorian imagination', *Literature and Medicine* **16(2)**, 141–173.

E. Gurney, F. W. H. Myers & F. Podmore (1886). *Phantasms of the Living*. Trübner & Co., London.

P. Haining (ed.) (1978). *The H. G. Wells Scrapbook*. New English Library, London.

J. C. Halimeh, T. Ergin, J. Mueller, N. Stenger & M. Wegener (2009). 'Photorealistic images of carpet cloaks', *Optics Express* **17**, 19328–19336.

J. C. Halimeh and M. Wegener (2012). 'Photorealistic ray tracing of free-space invisibility cloaks made of uniaxial dielectrics', *Optics Express* **20**, 28330–28340.

J. C. Halimeh and M. Wegener (2013). 'Photorealistic rendering of unidirectional free-space invisibility cloaks', *Optics Express* **21**, 9457–9472.

T. H. Hall (1962). *The Spiritualists: The Story of Florence Cook and William Crookes*. Duckworth, London.

P. M. Harman (2001). *The Natural Philosophy of James Clerk Maxwell*. Cambridge University Press, Cambridge.

P. Harrison (2011). 'Adam Smith and the history of the Invisible Hand', *Journal of the History of Ideas* **72**, 29–49.

J. Harvey (2007). *Photography and Spirit*. Reaktion, London.

P. M. Heimann (1972). 'The *Unseen Universe*: Physics and the philosophy of nature in Victorian Britain', *British Journal for the History of Science* **6**, 73–79.

J. Henry (1986). 'Occult qualities and the experimental philosophy: active principles in pre-Newtonian matter theory', *History of Science* **24**, 335–381.

J. Henry (2012). *Religion, Magic and the Origins of Science in Early Modern England*. Ashgate, Aldershot.

A. Hessenbruch (2000). 'Science as public sphere: X-rays between spiritualism and physics', in *Wissenschaft und Öffentlichkeit in Berlin, 1870–1930*, ed. C. Goschler. Franz Steiner Verlag, Stuttgart.

R. Hessler (1912). *Dusty Air and Ill Health*. Printed privately.

P. Holt (1992). 'H. G. Wells and the Ring of Gyges', *Science Fiction Studies* **19(2)**, 236–247. Available at http://www.depauw.edu/sfs/backissues/57/holt57art.htm.

R. Hooke (1665/2007). *Micrographia*. BiblioBazaar, Charleston, SC.

A. A. Hopkins (1898/1976). *Magic: Stage Illusions, Special Effects and Trick Photography*. Dover, New York.

J. C. Howell & J. B. Howell (2013). 'Simple, broadband, optical spatial cloaking of very large objects', preprint at http://www.arxiv.org/1306.0863.

X. Hu, C. T. Chan, K.-M. Ho & J. Zi (2011). 'Negative effective gravity in water waves by periodic resonator arrays', *Physical Review Letters* **106**, 174501.

J. Hughes (2003). 'Occultism and the atom: the curious history of isotopes', *Physics World*, September, 31–35.

K. Jacobs (1990). 'Freud, Blondlot, and the logic of invisibility', *Qui Parle* **4**, 21–46.

V.-M. Kärkkäinen (2002). *Pneumatology*. Baker Academic, Grand Rapids, MI.

D. S. Katz (2005). *The Occult Tradition: From the Renaissance to the Present Day*. Jonathan Cape, London.

C. Keller (ed.) (2009). *Brought To Light: Photography and the Invisible 1840–1900*. San Francisco Museum of Modern Art/Yale University Press, New Haven.

R. Kieckhefer (1990). *Magic in the Middle Ages*. Cambridge University Press, Cambridge.

R. Kipling (1920). 'Wireless', in *Traffics and Discoveries*. Doubleday, Page & Co., New York.

I. M. Klotz (1980). 'The N-Ray Affair', *Scientific American* **242**, 168–175.

Y. Lai, J. Ng, H. Chen, D. Han, J. Xiao, Z.-Q. Zhang & C. T. Chan (2009). 'Illusion optics: the optical transformation of an object into another object', *Physical Review Letters* **102**, 253902.

A. Lang (1894). *Cock-Lane and Common Sense*. Longmans, Green & Co., London.

C. Lavers (2008). 'Invisibility rules the waves', *Physics World* March, 21–25.

C. W. Leadbeater (1942). *Man Visible and Invisible*. Theosophical Publishing House, Madras.

W. Lefèvre (ed.) (2007). 'Inside the camera obscura: optics and art under the spell of the projected image.' Max Planck Institute for the History of Science preprint 333.

U. Leonhardt (2006). 'Optical conformal mapping', *Science* **312**, 1777–1780.

U. Leonhardt (2009). 'Towards invisibility in the visible', *Nature Materials* **8**, 537–538.

U. Leonhardt & T. G. Philbin (2009). 'Transformation optics and the geometry of light', *Progress in Optics* **53**, 69–152.

U. Leonhardt & T. Philbin (2010). *Geometry and Light: The Science of Invisibility*. Dover, Mineola.

U. Leonhardt (2011). 'What we won't be seeing', *Physics World* **24**(7), 26–29.

E. Lévi (1860/1922). *The History of Magic*, trans. A. E. Waite. William Rider & Son, London.

E. Lévi (1896), *The Doctrine of Transcendental Magic*, trans. A. E. Waite. Rider & Co., London.

J. Li & J. B. Pendry (2008). 'Hiding under the carpet: a new strategy for cloaking', *Physical Review Letters* **101**, 203901.

T. M. Lieber (1972). 'Ralph Ellison and the metaphor of invisibility in black literary tradition', *American Quarterly* **24**, 86–100.

J. London (1903). 'The Shadow and the Flash'. Available at http://www.eastoftheweb.com/short-stories/UBooks/ShaFla.shtml.

X. Luo, T. Yang, Y. Gu & H. Ma (2008). 'Conceal an entrance by means of superscatterer', preprint http://arxiv.org/abs/0809.1823.

S. L. MacGregor Mathers (1975) (ed.). *The Book of the Sacred Magic of Abra-Melin the Mage*. Dover, New York.

M. C. Malley (2001). *Radioactivity: A History of a Mysterious Science*. Oxford University Press, Oxford.

M. Mamei & F. Zambonelli (2003). 'Spray computers: frontiers of self-organization for pervasive computing'. Available at mars.ing.unimo. it/Zambonelli/PDF/woa03.pdf.

I. Marathakis (2007). 'From the Ring of Gyges to the Black Cat Bone: a historical survey of the invisibility spells.' Preprint, available at http:// www.hermetics.org/Invisibilitas.html.

J. Maskelyne (1949). *Magic: Top Secret*. Stanley Paul, London.

J. Matlock (1996). 'Reading invisibility', in M. B. Garber, P. R. Franklin & R. L. Walkowitz (eds), *Field Work: Sites in Literary and Cultural Studies*, Ch. 21. Routledge, New York.

M. McCall & P. Kinsler (2011). 'Cloaking space-time', *Physics World* **24**(7), 35–39.

C. McIntosh (1980). *The Rosy Cross Unveiled: The History, Mythology and Rituals of an Occult Order*. Aquarian Press, Wellingborough.

J. McIntyre (1997). *The Shape of Pneumatology*. T&T Clark, Edinburgh.

S. McLean (2009). *The Early Fiction of H. G. Wells: Fantasies of Science*. Palgrave Macmillan, London.

K. McMullen (dir.) (1983). *Ghost Dance* [film].

R. G. Medhurst & K. M. Goldney (eds) (1972). *Crookes and the Spirit World*. Souvenir Press, London.

C. Miéville (2011). *The City and the City*. Pan Macmillan, London.

E. P. Mitchell (1881). 'The Crystal Man'. Available at http://gutenberg.net. au/ebooks06/0602521h.html.

R. L. Moore (1977). *In Search of White Crows: Spiritualism, Parapsychology, and American Culture*. Oxford University Press, New York.

L. E. Morel (ed.) (2004). *Ralph Ellison and the Raft of Hope*. University Press of Kentucky, Lexington.

I. R. Morus (2005). *When Physics Became King*. University of Chicago Press, Chicago.

H. Murphy & M. Bellamy (2009). 'The dazzling zoologist: John Graham Kerr and the early development of ship camouflage', *The Northern Mariner* **19**, 171–192.

A. Nadel (1988). *Invisible Criticism: Ralph Ellison and the American Canon*. University of Iowa Press, Iowa City.

S. Natale (2011). 'A cosmology of invisible fluids: wireless, X-rays, and psychical research around 1900', *Canadian Journal of Communication* **36**, 263–275.

R. Noakes (1999). 'Telegraphy is an occult art: Cromwell Fleetwood Varley and the diffusion of electricity to the other world', *British Journal for the History of Science* **32**, 421–459.

R. Noakes (2004). 'The bridge which is between physical and psychical research: William Fletcher Barrett, sensitive flames, and spiritualism', *History of Science* **42**, 419–464.

R. Noakes (2007). 'Cromwell Varley FRS, electrical discharge and Victorian spiritualism', *Notes and Records of the Royal Society* **61**, 5–21.

R. Noakes (2008). 'The "world of the infinitely little": connecting physical and psychical realities circa 1900', *Studies in the History and Philosophy of Science* **39**, 323–334.

F. O'Brien (1881). 'What was it?'. In B. Matthews (ed.) (1907). *The Short-Story*, Ch. 13. American Book Company, New York. Available at http://www.bartleby.com/195/13.html.

Y. Oida & L. Marshall (1997). *The Invisible Actor*. Methuen, London.

W. J. Padilla, D. N. Basov & D. R. Smith (2006). 'Negative refractive index metamaterials', *Materials Today* **9(7–8)**, 28–35.

R. Panek (2005). *The Invisible Century: Einstein, Freud and the Search for Hidden Universes*. Harper, London.

G. Parkinson (2008). *Surrealism, Art and Modern Science*. Yale University Press, New Haven.

F. A. Pattie (1994). *Mesmer and Animal Magnetism*. Edmonston Publishing, Hamilton, NY.

J. B. Pendry & D. R. Smith (2004). 'Reversing light with negative refraction', *Physics Today* **57(6)**, 37–43.

J. B. Pendry, D. Schurig & D. R. Smith (2006). 'Controlling electromagnetic fields', *Science* **312**, 1780–1782.

R. Penrose (1981). *Picasso: His Life and Work*, 3rd edn. Granada, St Albans.

S. Perkowitz (2011). 'Deflecting the light and deceiving the gods', *Physics World* **24(7)**, 21–25.

J. D. Peters (1999). *Speaking Into the Air*. University of Chicago Press, Chicago.

R. Philmus (1970). *Into the Unknown: The Evolution of Science Fiction from Francis Godwin to H. G. Wells*. University of California Press, Berkeley.

Plato (4th century BC/2000). *The Republic*, trans. G. M. A. Grube. Pan, London.

R. Porter (ed.) (1996). *The Cambridge Illustrated History of Medicine*. Cambridge University Press, Cambridge.

C. Priest (2005). *The Glamour*. Gollancz, London.

S. Quinn (1995). *Marie Curie: A Life*. Simon & Schuster, New York.

R. Ralley (2010). *Magic: A Beginner's Guide*. OneWorld, Oxford.

N. Rankin (2008). *Churchill's Wizards: The British Genius for Deception 1914–1945*. Faber & Faber, London.

S. Richards (1982). *Invisibility: Mastering the Art of Vanishing*. Aquarian Press, Wellingborough.

J. R. Roach (1982). 'Garrick, the ghost and the machine', *Theatre Journal* **34(4)**, 431–440.

H. Robbins (2003). 'The Emperor's new critique', *New Literary History* **34**(4), 659–675.

J. Russell, B. Gee & C. Bullard (2012). 'Why do young children hide by closing their eyes? Self-visibility and the developing concept of self', *Journal of Cognition and Development* **13**(4), 550–576.

E. Schenkel & S. Welz (eds) (2007). *Magical Objects: Things and Beyond*. Galda & Wilch Verlag, Berlin.

D. Schurig, J. J. Mock & D. R. Smith (2006). 'Electric-field-coupled resonators for negative permittivity metamaterials', *Applied Physics Letters* **88**, 1–3.

D. Schurig, J. J. Mock, B. J. Justice, S. A. Cummer, J. B. Pendry, A. F. Starr & D. R. Smith (2006). 'Metamaterial electromagnetic cloak at microwave frequencies', *Science* **314**, 977–980.

J. Sconce (2000). *Haunted Media: Electronic Presence from Telegraphy to Television*. Duke University Press, Durham.

R. Scot (1584/1972). *Discoverie of Witchcraft*. Dover, New York.

H. L. Shaw (1977). *Hypnosis in Practice*. Baillière Tindall, London.

R. A. Shelby, D. R. Smith, S. C. Nemat-Nasser & S. Schultz (2001). 'Microwave transmission through a two-dimensional, isotropic, left-handed metamaterial', *Applied Physics Letters* **78**, 489–491.

R. A. Shelby, D. R. Smith & S. Schultz (2001). 'Experimental verification of a negative index of refraction', Science 292 no. 5514 (2001), pp.77–79.

B. Sidis (1921). *Psychology of Suggestion*. D. Appleton & Co., New York.

E. Silby (1795). *A New and Complete Illustration of the Occult Sciences*. C. Stalker, London.

D. R. Smith & N. Kroll (2000). 'Negative refractive index in left-handed materials', *Physical Review Letters* **85**, 2933–2936.

D. R. Smith, W. J. Padilla, D. C. Vier, S. C. Nemat-Nasser & S. Schultz (2000). 'Composite medium with simultaneously negative permeability and permittivity', *Physical Review Letters* **84**, 4184–4187.

D. R. Smith & D. Schurig (2003). 'Electromagnetic wave propagation in media with indefinite permittivity and permeability tensors', *Physical Review Letters* **90**, 077405.

D. R. Smith, J. B. Pendry & M. C. K. Wiltshire (2004). 'Metamaterials and negative refractive index', *Science* **305**, 788–792.

D. R. Smith (undated). 'Metamaterials – History', http://people.ee.duke.edu/~drsmith/metamaterials/metamaterials_history.htm.

M. Solomon (2010). *Disappearing Tricks: Silent Film, Houdini, and the New Magic of the Twentieth Century*. University of Illinois Press, Urbana & Chicago.

B. M. Stafford (1994). *Artful Science*. MIT Press, Cambridge, MA.

B. M. Stafford & F. Terpak (2001). *Devices of Wonder*. Getty Publications, Los Angeles.

J. B. Stallo (1881). *The Concepts and Theories of Modern Physics*. Appleton & Co., New York.

G. Stein (1993). *The Sorcerer of Kings*. Prometheus, Buffalo, NY.

N. Stenger, M. Wilhelm & M. Wegener (2012). 'Experiments on elastic cloaking in thin plates', *Physical Review Letters* **108**, 014301.

H. Stephen (1885). 'No. 11 Welham Square', *Cornhill Magazine*, May. Available at http://gutenberg.net.au/ebooks06/0606521h.html.

M. Stephens & S. Merilaita (2009). 'Animal camouflage: current issues and new perspectives', *Philosophical Transactions of the Royal Society B* **364**, 423–427.

B. Stewart & P. G. Tait (1876). *The Unseen Universe, or Physical Speculations on a Future State*. Macmillan, New York.

A. Still (1944). *Soul of Amber*. Murray Hill Books, New York.

A. Still (1946). *Soul of Lodestone*. Murray Hill Books, New York.

C. Stryker (2012). *Hacking the Future*. Overlook Duckworth, New York.

L. E. Sullivan (ed.) (1987). *Hidden Truths: Magic, Alchemy, and the Occult*. Macmillan, New York.

D. Suvin & R. M. Philmus (eds) (1977). *H. G. Wells and Modern Science Fiction*. Buckwell University Press/Associated University Presses, Cranbury, NJ.

S. Tachi (2003). 'Telexistence and retro-reflective projection technology (RPT)', *Proceedings of the 5th Virtual Reality International Conference (VRIC2003)*, Laval Virtual, France, pp.69/1–69/9.

N. Taillepied (1600/undated). *A Treatise of Ghosts*, intro. M. Summers. Fortune Press, London.

M. Tatar (2008). *The Annotated Hans Christian Andersen*. W. W. Norton, New York.

K. Thomas (1991). *Religion and the Decline of Magic*. Penguin, London.

L. Thorndike (1934/1941). *A History of Magic and Experimental Science* Vols III–VI. Columbia University Press, New York.

J. Valentine, S. Zhang, T. Zentgraf, E. Ulin-Avila, D. A. Genov, G. Bartal & X. Zhang (2008) 'Three-dimensional optical metamaterial with a negative refractive index', *Nature* **455**, 376–379.

F. Vaz da Silva (2002). 'Vision beyond eyesight', *Separata dos Trabalhos de Antropologie e Etnologia* Vol. 42(1–2), 33–47. Porto.

B. Vickers (ed.) (1984). *Occult and Scientific Mentalities in the Renaissance*. Cambridge University Press, Cambridge.

H. A. Vinson (2011). *The Time Machine and Heart of Darkness: H. G. Wells, Joseph Conrad, and the fin de siècle*. Graduate thesis, University of South Florida.

J. Waller (2002). *The Discovery of the Germ*. Icon, London.

A. M. Walmsley (1967). *Anton Mesmer*. Robert Hale, London.

C. Webster (1974). 'New light on the Invisible College: the social relations of English science in the mid-seventeenth century', *Transactions of the Royal Historical Society* **24**, 19–42.

C. Webster (1982). *From Paracelsus to Newton: Magic and the Making of Modern Science*. Cambridge University Press, Cambridge.

H. G. Wells (1897/2005). *The Invisible Man*, ed. P. Parrinder, intro. C. Priest. Penguin, London.

R. H. West (1969). *The Invisible World: A Study of Pneumatology in Elizabethan Drama*. Octagon Books, New York.

L. White (2007). 'Damien Hirst, Colley Cibber and the bathos of the commercialised sublime', a paper given at the conference *Taste, Vision, Transcendence: Sublimity 1700–1900*, University of Sussex, 5th January 2007.

F. Wilczek (2013). 'Multiversality', preprint at http://www.arxiv.org/abs/1307.7376.

O. Wilde (1887). 'The Canterbury Ghost', in G. F. Maine (ed.) (1961), *The Works of Oscar Wilde*. Collins, London.

K. Williams (2007). *H. G. Wells: Modernity and the Movies*. Liverpool University Press, Liverpool.

C. Wilson (1995). *The Invisible World*. Princeton University Press, Princeton.

J. D. Wilson (1967). *What Happens in Hamlet*. Cambridge University Press, Cambridge.

T. Yang, H. Chen, X. Luo & H. Ma (2008). 'Superscatterer: enhancement of scattering with complementary media', preprint at http://arxiv.org/abs/0807.5038.

J. Yao, Z. Liu, Y. Liu, Y. Wang, C. Sun, G. Bartal, A. M. Stacy & X. Zhang (2008). 'Optical negative refraction in bulk metamaterials of nanowires', *Science* **321**, 930.

F. Zambonelli, M.-P. Gleizes, M. Mamei & R. Tolksdorf (2005). 'Spray computers: explorations in self-organization', *Pervasive and Mobile Computing* **1**, 1–20. Available at http://citeseerx.ist.psu.edu/viewdoc/download?doi=10.1.1.368.3167&rep=rep1&type=pdf.

F. Zambonelli & M. Mamei (2002). 'The cloak of invisibility: challenges and applications', *Pervasive Computing* **1**(4), 62–70. Available at http://www.agentgroup.unimo.it/didattica/cas/L17/Zambonelli_Invisibility.pdf.

B. Zhang, Y. Luo, X. Liu & G. Barbastathis (2011). 'Macroscopic invisibility cloak for visible light', *Physical Review Letters* **106**, 033901.

J. Zipes (2005). *Hans Christian Andersen: The Misunderstood Storyteller*. Routledge, New York.

J. Zipes (2012). *The Irresistible Fairy Tale*. Princeton University Press, Princeton.

Index